Cambridge Elements ≡

Elements in the Philosophy of Biology
edited by
Grant Ramsey
KU Leuven
Michael Ruse
Florida State University

ECOLOGICAL MODELS

Jay Odenbaugh
Lewis & Clark College

CAMBRIDGE
UNIVERSITY PRESS

CAMBRIDGE
UNIVERSITY PRESS

University Printing House, Cambridge CB2 8BS, United Kingdom

One Liberty Plaza, 20th Floor, New York, NY 10006, USA

477 Williamstown Road, Port Melbourne, VIC 3207, Australia

314–321, 3rd Floor, Plot 3, Splendor Forum, Jasola District Centre, New Delhi – 110025, India

79 Anson Road, #06–04/06, Singapore 079906

Cambridge University Press is part of the University of Cambridge.

It furthers the University's mission by disseminating knowledge in the pursuit of education, learning, and research at the highest international levels of excellence.

www.cambridge.org
Information on this title: www.cambridge.org/9781108728690
DOI: 10.1017/9781108685283

First published 2019

A catalogue record for this publication is available from the British Library.

ISBN 978-1-108-72869-0 Paperback
ISSN 2515-1126 (online)
ISSN 2515-1118 (print)

Ecological Models

Elements in the Philosophy of Biology

DOI: 10.1017/9781108685283
First published online: September 2019

Jay Odenbaugh
Lewis & Clark College

Abstract: In this Element, we consider three questions. What are ecological models? How are they tested? How do ecological models inform environmental policy and politics? Through several case studies, we see how these representations, which idealize and abstract, can be used to explain and predict complicated ecological systems. Additionally, we see how they bear on environmental policy and politics.

Keywords: models ecology, representation, environment, idealization, prediction

ISBNs: 9781108728690 (PB) 9781108685283 (OC)
ISSNs: 2515-1126 (online) 2515-1118 (print)

Contents

Introduction

In this short Element, I consider three questions.

- What are ecological models?
- How are they tested?
- How do ecological models inform environmental policy and politics?

Ecology is a remarkably diverse discipline since it includes physiological, behavioral, population, community, landscape, ecosystem, and evolutionary ecology. It frankly raises the question: In what sense are they all ecology (Cooper, 2003, ix–xiv)? For example, one might argue that ecosystem ecology is really just biogeochemistry – it isn't biology at all. Due to space, my examples of models can only pull from a few of these areas, but I think they are representative of the work that goes on in the science of the struggle for existence. Additionally, in order to understand contemporary ecological models, you must understand many models that appeared in the 1920s through to the 1990s (Kingsland, 1995). So, I pay special attention to this period although I include a variety of more recent models as well.

Ecological models are very often mathematical. I have tried to do justice to the models, which means including some of the mathematics. But, I have also tried to circumnavigate the math when I can. The tool I use to do so are *Technical Discussions*. Therein, I add details that can be skipped if you don't want them. I expect my reader to be either the ecologically curious philosopher or philosophically curious ecologist. Both constituencies are interested in learning about the issues that animate the other, even if I cannot fully explore those issues and thus satisfy those curiosities. Ecologists will find the models familiar, but hopefully the philosophical discussions will aid them in critical reflecting on their scientific practice. Philosophers will find the philosophy familiar, but the models less so and hopefully they push those conceptual issues further in new directions.

To warm you up, consider this. Models involve idealizations, simplifications, distortions of the truth, and the like. If science is searching for the truth, you would think they have no place in sciences like ecology. But they *do*. That is what this Element is about.

1 What Are Ecological Models?

If you leaf through any recent issue of *Ecology, Ecology Letters, Trends in Ecology and Evolution, Journal of Ecology*, among others, you will find reams of models.[1] Ecology includes a lot more than model-building, but it is central to the science. For example, modelers Sarah Otto and Troy Day analyzed the content of the journals *Evolution, Ecology,* and *American Naturalist* for the year 2001 (Otto and Day, 2011, 1–2). With regard to *Ecology,* they found that, of 274 articles, 100 percent used models if we include phylogenetic and statistical ones, 35 percent used models to predict results, and 33 percent explicitly present the model equations. That is a lot of models.

Increasing attention has been given by philosophers to the role of models and modeling in science (Bailer-Jones, 2009; Pincock, 2011). Here are two issues concerning the nature of models that have occupied them. First, the term "model" is applied to many different things, including mathematical structures, graphs, computer simulations, and organisms. What are models? Second, in some sciences we talk about theories, but in others we talk of models. This raises the question: How are models and theories related? Ecologists like Richard Levins claim theories just are a "cluster of models" (Levins, 1966, 431). In this section, we will consider both of these questions.

1.1 Metapopulation Models

In order to help us think about what ecological models are, let's begin with an example used throughout the section.[2] Let's consider some metapopulation models from population ecology (Levins, 1969; Levins and Culver, 1971; Lande, 1987; Gotelli, 1991).[3] A metapopulation can be thought of as a "population of populations" that are subdivided spatially into patches, but are connected by immigrating and emigrating organisms. Let P be the proportion of occupied population patches where $0 \leq P \leq 1$. If $P = 1$, then all patches are occupied, and if $P = 0$, then extinction (at least regionally) has occurred. Thus, $(1 - P)$ is the proportion of unoccupied patches. All metapopulation models have this form,

$$\frac{dP}{dt} = \text{immigration rate} - \text{extinction rate}$$

Let c be probability of local colonization, and e be the probability of local extinction.[4] The simplest metapopulation model assumes an "island–mainland"

---

[1] And note, these are not journals like *Theoretical Ecology* and *Theoretical Population Biology* where you expect a lot of models.

[2] Due to space, I focus on mathematical models. However, ecology also uses material models as well – see Griesemer (1990a, 1990b); Odenbaugh (2010), Weisberg (2012) for further discussion.

[3] For a presentation of metapopulation models, see Gotelli (1995, ch. 4) and Rockwood (2015, ch. 5).

[4] By local probability, we are considering the probability of colonization or extinction of a patch. We will then consider the immigration or extinctions rates more generally.

structure of immigration. Colonists always come to a patch from some external source sometimes called a "propagule rain." Let's also assume that the extinction rate is independent of regional occurrence of the species. That is, the probability of extinction is wholly unaffected by how many patches are occupied. We then have,

$$\frac{dP}{dt} = c(1 - P) - eP \tag{1.1}$$

Next, let's suppose our metapopulation is not changing; it is at an equilibrium, and so $dP/dt = 0$.

TECHNICAL DISCUSSION

Just because a system is at equilibrium does not imply it is a *stable* equilibrium. Suppose we have a model of the form,

$$\frac{dP}{dt} = F(P)$$

We find the equilibria by setting $dP/dt = 0$ and solving for $\hat{P}$. But, we haven't determined if the equilibrium is stable. To determine if the equilibrium is stable, we define $P = \hat{P} + p$ and find $dp/dt \approx \lambda p$, which is,

$$\lambda = \left.\frac{dF}{dP}\right|_{P=\hat{P}}$$

The equilibrium is stable if λ, is negative and it is unstable if λ is positive (Hastings, 1997, 91). The rate of return (or away) is determined by λ. Things are more complicated when the model has two or more variables of course.

The equilibrium metapopulation $\hat{P}$ is,

$$\hat{P} = \frac{c}{c + e}$$

Even when c is small and e is large, $P > 0$. Thus, the metapopulation always persists since the immigration rate is always positive given immigrants are always entering from the source.

We can revise our model (1.1) by assuming internal colonization. This means colonists come from other patches rather than a mainland. There is no external source, only an internal one. Specifically, we assume that when P is large, the immigration rate is low because there are few places to immigrate. But, when P is small, the immigration rate is also low because there are few sites from which colonists can be found. We thus alter the immigration term from $c(1 - P)$ to $cP(1 - P)$. Thus, we have,

$$\frac{dP}{dt} = cP(1 - P) - eP \qquad (1.2)$$

At equilibrium,

$$\hat{P} = 1 - \frac{e}{c}$$

The metapopulation avoids extinction when the immigration rate is greater than the extinction rate; otherwise, it goes extinct. It also entails that, if the rate of extinction is non-zero, then the there must be habitats unoccupied.

Both (1.1) and (1.2) assume that the extinction rate is independent of how many patches are occupied. But, we might think immigrants could prevent a patch from extinction. We thus replace eP with $eP(1 - P)$. This is termed the "rescue effect." Thus, (1.1) can be amended with a propagule rain and rescue effect,

$$\frac{dP}{dt} = c(1 - P) - eP(1 - P) \qquad (1.3)$$

At equilibrium, we have,

$$\hat{P} = \frac{c}{e}$$

As with (1.1), the persistence of the metapopulation is assured and all patches are occupied when $c > e$. Finally, we can suppose immigration and extinction are dependent on regional occurrence,

$$\frac{dP}{dt} = cP(1 - P) - eP(1 - P) \qquad (1.4)$$

With (1.4), there is no simple equilibrium. If $c > e$, then $cP(1-P) > eP(1-P)$ and the metapopulation increases until all patches are occupied. If $c < e$, then $cP(1-P) < eP(1-P)$ and the metapopulation decreases until extinct. If $c = e$, then we have a neutral equilibrium.[5]

As one more final refinement, consider the work of Russell Lande (1987). Suppose a fraction $(1 - h)$ of patches are unsuitable, and thus cannot be colonized. Thus, h is the fraction that are suitable. Therefore, the colonization rate of empty patches is $cPh(1 - P)$. Assuming internal immigration and no rescue effect, we have,

$$\frac{dP}{dt} = cPh(1 - P) - eP \qquad (1.5)$$

---

[5] A neutral equilibrium is one where, if the system is displaced from x^*, it does not return to x^*, but remains at the new equilibrium.

At equilibrium,

$$\hat{P} = 1 - \frac{e}{ch}$$

Thus, $\hat{P} > 0$, if $1 > e/ch$ and $h > e/c$.

TECHNICAL DISCUSSION

Let P^* be the fraction of habitat originally occupied and h_0 be the original amount of suitable habitat. Our equilibrium for (1.5) is,

$$0 = ch(1 - P) - e$$

We can substitute these values in the equation and rearrange the terms,

$$\frac{e}{c} = h_0(1 - P^*)$$

Since the metapopulation can persist if $h > e/c$,

$$h > h_0(1 - P^*)$$

This is termed "Levins's rule." Metapopulation survival occurs if the remaining number of patches following habitat reduction is greater than the number of empty although suitable patches prior to the reduction.

Carlson (2000) applied Levins's rule to the woodpecker *Dendrocopos leucotos* in Sweden and Finland. He determed that $h_0 = 0.66$ and $P^* = 0.81$. Thus, $h > 0.125$ for the metapopulation to survive. However, he determined $h < 0.12$, which was confirmed by the fact that the populations have been declining rapidly.

Our simple metapopulation models are based on a variety of assumptions. Here are some noteworthy ones.

- The probabilities of immigration and extinction are influenced by the number of occupied patches, but not by their spatial arrangement.
- The local probabilities of immigration and extinction are constants since they do not change with time.
- There are a large number of homogeneous patches.

For actual metapopulations, they are all probably *false*. Spatial arrangement clearly matters, since, the closer an occupied patch is to an unoccupied one, the more likely colonization will occur. Probabilities of colonization and extinction surely change over time. Additionally, the patches differ in their quality, which affects extinction rates.

1.2 What Are Models?

Ecologists use the term "model" in myriad ways, as mentioned earlier. Mathematical structures, graphs, computer simulations, organisms, among others, are all called 'models.' One might infer there is no unified account of what models are, but, we can resist this inference. Here is a working definition of the term:

> Models are representations that abstract and idealize.

Let's unpack it.

First, models have intentional content – they are about what they represent. There are many different kinds of representations; for example, words, pictures, and so on. Models represent their objects by attributing properties to them – by representing them as being a certain way. A model may have an intension, but no extension (or no referent). The intentional content of models can be understood in terms of accuracy conditions – the content of a model is the way the world would have to be for the model to have a referent. A model has this content even in cases in which the world isn't this way, and thus fails to have a referent. The accuracy conditions of ecological models are often truth conditions, since they are equations that describe ecological systems truly (or not).[6]

Second, models represent their objects by abstraction: they represent their objects by attributing to them a proper subset of the properties that may be so attributed. Suppose I draw a picture of my son Everett, niece Sadie, and nephews Caleb and Jack using a graphite pencil. My drawing does not represent Everett, Sadie, Caleb, and Jack as monochromatic. Rather, my drawing simply does not represent the colors of their clothes, hair, and so on.

Third, models represent their objects by idealization: they represent their objects by attributing to them properties they lack (think of a caricaturist's drawings). Lots of representations misrepresent their referents, of course. The history of science is chock full of false theories and models. But models are not merely false. Their idealizations are also *useful* (and they might even be approximately true).[7] Successful representation sometimes requires the attribution of

---

[6] Truth conditions are a subset of accuracy conditions. For example, pictures may be more or less accurate, but we might not think they express propositions that are true or false. For more on depiction, see Kulvicki (2013).

[7] Idealizations can be useful for one purpose and not for another. Really simple population growth equations in population ecology or single locus, two allele models in population genetics are often of little use predictively. But, they are of great value pedagogically, since they help students learn how to build and evaluate models. Anchoring model evaluation to the purposes for which they are built or deployed is crucial (Odenbaugh, 2005; Parker, 2009).

properties not had by the object represented.[8] Thus, the intension of the term 'model' is abstract, idealized representation. The extension of the term, however, consists in a variety of things that have those properties. That is, objects as diverse as phase spaces, computer code in R, and *Tribolium castaneum* are all objects denoted by the term 'model'.

Our working definition fits metapopulation models very well. First, our models are representations – they represent populations of populations as having various properties like habitat patches where they reside, probabilities of colonization and extinction, and so on. Second, the models abstract from those metapopulations. For example, organisms in those populations compete with one another and our models are silent on this score. They do not say that they don't compete. Rather, they simply do not describe intraspecific competition at all. Third, the models idealize those metapopulations too. They assume that patches are homogeneous, the spatial distribution of patches does not matter, the probabilities of local colonization and extinction are constants, among others. These are all false assumptions, but which provide us with useful analytically tractable equations for thinking about spatial ecology.

Modelers talk of models being built from "assumptions." Sometimes they talk of models "assuming" this and that. Assumptions are the propositions that characterize the model. Mathematician Edward Bender puts it this way,

> Definitions of the variables and their interrelations constitute the *assumptions* of the model. We then use the model to *draw conclusions* (i.e., to make predictions). This is a deductive process: *If the assumptions are true, the conclusions must also be true.* Hence a false prediction implies that the model is wrong in some respect. (Bender, 1978, 4)

This is a shorthand for talking about the structure of models (see Sorensen, 2012).

In this Element, we will discuss mostly mathematical models. Modelers describe models as having variables, parameters, and laws. The variables of a model are those properties that can change. In our metapopulation models, the variable P is the frequency of occupied patches. The parameters of the models are those properties that cannot, or at least do not, change. The parameters in the metapopulation models are c, e, and h. Of course, in the actual world, these

---

[8] For discussions of idealization and abstraction see (McMullin, 1985; Cartwright, 1994; Morrison, 2015; Potochnik, 2017; Weisberg, 2012). One might argue that successful models abstract but rarely idealize (Strevens, 2008). For example, it is common in evolutionary genetics to suppose we have an infinite population size that would clearly be an idealization ignoring random genetic drift. However, we could also describe this as assuming a sufficiently large population such that drift can be ignored (Strevens, 2017). For our purposes, we will assume models in ecology abstract and idealize.

properties might change over time, but, in our model, we simplify by supposing they don't. Finally, the laws of the model are laws of "succession" or "coexistence." The former tell us how a system changes over time. For example, $dP = c(1 - P) - eP$ is a law of succession. Laws of succession can be deterministic or stochastic. If deterministic, then, given values of c, e, and h and P at a time, there is there is a one value of P possible for any given time. Similarly, if stochastic, then, given the same conditions at a time, it is possible for P to have more than one value at any later time. Likewise, they can be continuous or discrete. A function is continuous with respect to any interval if it can take any value in the interval; otherwise, it is discrete.[9] The laws of coexistence tell us what values of the properties (e.g., variables and parameters) can be jointly occupied by the system. For example, the equilibrium $\hat{P} = c/(c + e)$ is such a law.

This notion of a 'law' is not the same as discussed by philosophers of science and metaphysicians. First, the laws in ecologist's models are simply generalizations. They need not be exceptionless generalizations that support counterfactuals, as philosophers sometimes put it. And, they may be false. Second, philosophers of biology are divided over what laws are and whether biology has them. Some argue biology has no such laws since biological generalizations are mathematical truisms (Sober, 1997), true counterfactual-supporting generalizations that have exceptions (Brandon, 1997), or that they are the contingent products of evolutionary history and false in some circumstances (Beatty, 1997). Others argue that physical and chemical generalizations have exceptions and are contingent; hence, physics, chemistry, and biology are in the same boat (Cartwright, 1983).[10]

A very popular view of models is what I call the "similarity view" (Hesse, 1966; Giere, 1988; Weisberg, 2012). This view says that a representational device represents an object (if it does) that is similar in certain degrees and respects to the object. In the case of mathematical models, we have a set of equations that refer to a mathematical structure. This mathematical structure when interpreted is then similar in certain respects in certain degrees to the object. It it this interpreted mathematical structure that is the model.[11] Thus,

---

[9] It is worth noting that this is an idealization in ecology, since population size is not continuous. But, for some populations, this is approximately true.

[10] In ecology, there is rich debate over ecological laws among philosophers and ecologists (Cooper, 1998; Lawton, 1999; Weber, 1999; Turchin, 2001; Berryman, 2003; Colyvan and Ginzburg, 2003; Mikkelson, 2003; Lange, 2005; Linquist et al., 2016).

[11] The similarity view is related to the semantic view of theories (Beatty, 1980; van Fraassen, 1980; Beatty, 1982; Suppe, 1989; Thompson, 1989; Lloyd, 1994). For an analysis of the equilibrium model of island biogeography using the semantic view of theories, see Castle (2001). For an

model (1.1) refers to the unit interval $[0, 1] = \{x \mid 0 \leq x \leq 1\}$. This mathematical structure may be similar to the Bay checkerspot butterfly's (*Euphydryas editha bayensis*) patch occupancy (Harrison et al., 1988). This butterfly lives in discrete patches that are organized in a metapopulation. Adult butterflies appear in the spring and females lay their eggs on *Plantago erecta*. This host provides food for capterpillars, which feed for a few weeks, and then go into a summer diapause. They return to feeding in December until February, and then build cocoons. *P. erecta* live in Northern California on soil rock outcroppings. The butterfly and host can go out of synchrony with extreme weather like droughts, which lead to local extinctions. However, other patches provide available colonists to immigrate to new hosts in the old patches.

The similarity view faces a problem, which I call "Hughes' worry" (Odenbaugh, 2015, 2018). This problem was first posed by R. I. G. Hughes (1997). Consider our model (1.1) again, $dP/dt = c(1 - P) - eP$. The variable P represents the proportion of patches occuped, c is the probability of local immigration, and e is the probability of local extinction. An object can have a probability of immigration or extinction only if it can immigrate or go extinct. Mathematical objects like real numbers certainly cannot immigrate or go extinct. So, they cannot have properties like a probability to immigrate or go extinct (even if they can have probabilities). Mathematical objects and metapopulations cannot share the properties like *probability to immigrate* and *probability to go extinct*. Therefore, they cannot be similar with respect to those properties. But this implies that the similarity view is incorrect.

One response is that we should we think of the data (e.g., the metapopulation) as a mathematical structure too. If right, then, certainly, mathematical structures can share properties. For example, the numbers 2, 3, 5, 7, 11, 13, 17, 19, 23 and 29 all have the property of *being prime*, but we still have the same problem we started with – how can can the data structure be similar to the metapopulation?[12]

My own view of models is a deflationary one (Downes, 1992; Callender and Cohen, 2006; Suárez, 2010, 2015). Many philosophers of science think

---

overview of the received view, the semantic view, and models as mediators in the context of biology, see Odenbaugh (2010).

[12] Here is one way to represent data mathematically. A relational structure is a set of objects D with relations R on them; $M = \langle D, R \rangle$. Let that be our model. Additionally, suppose the data as a relational structure $M^* = \langle D^*, R_i^* \rangle$. There might be an isomorphism between M and M^* that is a function f such that $\langle o_1, \ldots, o_n \rangle \in R$ if, and only if, $f(o_1), \ldots, f(o_n) \rangle \in R^*$. But, at best, we have shown that there is a second-order relation of isomorphism between the two relational structures. There is no R or R^*, respectively such that an element of D and D^* both have it. Therefore, even if here are mappings between interpreted relational structures, these are not the relevant shared spatiotemporal properties between mathematical and concrete objects.

a special account of representation is needed for models (Giere, 1988; Hughes, 1997; van Fraassen, 2010). This isn't wrong *per se*. However, if the above argument is sound, then it is unnecessary. Following Callender and Cohen (2006), there are fundamental and derived representations (Grice, 1991). The latter are explained in terms of the former. For example, the meaning of utterances and inscriptions are explained in terms of mental states, which in turn are explained in terms of something more fundamental (insert your favorite naturalistic theory of mental representation). Deflationists deny we need a tailored account of scientific representation for models. We simply deploy those general accounts of representations found in cognitive science, cognitive psychology, linguistics, and so on (Cummins, 1989; Sterelny, 1990). This is consistent with there being different types of representation, which there are of course. For example, pictures and words represent in distinctive ways (Goodman, 1968), but these differences are orthogonal to the model/non-model distinction.

The similarity view departs from deflationism in two ways. First, it supposes that models are a special *sui generis* form of indirect representation.[13] Second, the representations are not true of or satisfied by objects, but are aspectually similar to the represented. We don't need to make either supposition.

Why accept deflationism? First, scientific representations are constructed from ordinary representational tools like languages, diagrams, among others. Second, the features that make scientific representation seem distinct from other forms of representation are actually found in them too. Models are representations that involve abstraction and idealization, but these are found in other types of representation. As we saw, a graphite pencil drawing does not represent colors of objects. In language, we presuppose sharp boundaries where there are none between things. Third, philosophers like Giere (1988, 1999, 2010), Hughes (1997), and van Fraassen (2010) already employ a deflationary framework construing representation in terms of intentions and interpretation.

As is customary in cognitive science and the philosophy of mind, we can distinguish between representational vehicles and representational contents (Dretske, 1997). Representational vehicles are the objects, events, or properties that do the representing. Representational contents are the properties the vehicles represent objects as having. Scientists use various vehicles to represent the world, including concrete objects, equations, graphs, pictures, and so on

---

[13] A representation x is indirect if x represents y, which in turn represents z rather than x directly representing z. Representation is not a transitive relation, and so x does not represent z. On the similarity view, equations represent mathematical structures that represent and are similar to ecological systems. For deflationists, the equations directly represent systems, ecological and otherwise. Although it is common, it is a misconception to think that similiarists define representation in terms of similarity.

(Perini, 2005a, 2005b). These representational vehicles and the content they express *are* the models. When we "write down a model" we are doing just that. For example, if I say the following in a theoretical ecology class,

> Let '*P*' be a population's occupancy of patches, '*c*' be its probability of local colonization, and '*e*' be its probability of local extinction. And, let $dP/dt = c(1 - P) - eP$ describe how patch occupancy changes over time.

then I have given you a model of a metapopulation. On a deflationary view, (1.1) is the model and it represents metapopulations.[14] This avoids Hughes' worry since models do not describe mathematical objects.[15] We can also make good sense of how models relate to the world – they do so just like other representations. They are true or false of it. This avoids confusing properties of vehicles and contents.

I now want to turn to the relationship between models and theories.

1.3 Models and Theory

If ecology, or biology more generally, lacks laws traditionally construed, and a scientific theory consists in such laws, then, apparently, ecology also lacks scientific theories (Smart, 1963). As noted earlier, there is a different way of thinking about scientific theories. From this view, theories are clusters or families of models (Giere, 1988, 1999, 2010). Let's consider our metapopulation models again.

The metapopulation models we considered all represent populations of populations. Additionally, they are related to one another. Consider model (1.5) $dP/dt = cPh(1 - P) - eP$. In this model, we assumed $(1 - h)$ is the fraction of patches that are permanently destroyed and cannot be colonized, and thus h is the fraction that can be colonized. If $h = 1$, then we just have Levins's original model (1.2) $dP/dt = cP(1 - P) - eP$. As another example, note that as $P \to 1$, then $cP(1 - P)$ reduces to $c(1 - P)$. Thus, for large values of P, (1.1) and (1.2) are approximately the same. Last, as $P \to 0$, then $eP(1 - P)$ reduces to eP. Thus, our metapopulation models are not an arbitrary set of models

---

[14] To say that the classic Levins's model represents metapopulations of course is not to say that it *accurately* does. Accurate representation will be the subject of the next section.

[15] Although ecological models describe ecological systems, there may be parts of them that do not refer ecological properties. For example, stability analyses of ecological models sometimes involve complex numbers like $i = \sqrt{-1}$, which are the sum of an imaginary number and a real number (Hastings, 1997, 158). It is plausible to say that P, r, q and h refer to ecological objects and properties. However, i seems to refer to a pure mathematical structure. How we should understand imaginary numbers is a problem in mathematics and the philosophy of mathematics that we can thankfully leave to the side.

representing a shared domain, but are logically related to one another.[16] So, how are models able to form a family? We start with a basic model (1.1) from which all the others are generated. This model forms the basis of a structural framework, from which other models "descend" by virtue of having different assumptions.[17] In metapopulation theory, we have a family of models starting with the assumption of a propagule rain and no rescue effect. We then can add internal colonization or the rescue effect. Additionally, we can also add the assumption that some patches are permanently destroyed and so forth.[18]

Ecological systems consist in a large number of interacting processes. Some have argued that this fact coupled with our own cognitive limitations and resources requires that we must use multiple models.[19] First, the parameters and variables cannot be measured in real time with so few field or laboratory biologists. Second, a single "true equation" would be incredibly difficult to understand and use given it consists in so many partial differential equations with hundreds of parameters and variables. Third, the resulting equations will be analytically insoluble. If ecologists aim to have models that represent biological systems realistically, generally, and precisely, then there will be tradeoffs with respect to these model desiderata (Levins, 1966). Some of these difficulties can be mitigated. For example, even insoluble equations can be simulated numerically on computers, providing an inductive survey of their dynamical behavior (Bernstein, 2003). That is, we examine our model's dynamical behavior by sampling regions of parameter space that are of interest. Thus, this debate is still ongoing.

Another important technique for analyzing clusters of models is robustness analysis. Here is how mathematician Edward Bender describes it,

> A result is *robust* if it can be derived from a variety of different models of the same situation, or from a rather general model. A prediction that depends on very special assumptions for its validity is *fragile*. The cruder the model, the less believable its fragile predictions. (Bender, 1978, 4)

For example, (1.2) and (1.5) predict very similar equilibria if $h \approx 1$, since $e/c \approx e/ch$ (see Etienne, 2002).[20] There is an interesting question as to what

---

[16] In fact, one can devise a general model from which all of our metapopulation models can be derived (Gotelli and Kelley, 1993).

[17] The evolutionary metaphor is imperfect since we have homologies and analogies as well. But, we can see how theories might be families of models.

[18] For an extension of (1.2) to two species see Horn and MacArthur (1972).

[19] See Levins (1966); Odenbaugh (2003, 2006a); Wimsatt (2007); Matthewson and Weisberg (2009); Matthewson (2011); Weisberg (2012), although also Orzack and Sober (1993); Orzack (2005).

[20] Robustness analysis is much simpler when we have nested models as opposed to non-nested models. A model is nested in another if the former is a special case of the latter (Hilborn and

makes models 'different' and of the 'same situation.' First, models must represent the same properties. One of the ways this can occur is if they have the same dependent variables. Second, models are different if they make different predictions. Thus, (1.2) and (1.5) are different since they make different predictions for some values of the variable and parameters. If the models are different but make the same predictions, then we need to determine that their assumptions are independent of one another. That is, they do not have the same truth conditions or are independent in the statistical sense (i.e., the events or random variables they describe are independent).[21] If this can be done, we have, as Richard Levins describes it,

> Therefore, we attempt to treat the same problem with several alternative models, each with different simplifications, but with a common biological assumption. Then, if these models, despite their different assumptions, lead to similar results we have what we can call a robust theorem that is relatively free of the details of the model. Hence, our truth is intersection of independent lies. (Levins, 1966, 423)

But, you might naturally ask: What is a "true lie?"

1.4 True Lies and Approximate Truth

Philosophers of science have routinely assumed, and sometimes argued, that if a theory or model explains some explanandum, then it must be true (Hempel, 1965a; Salmon, 1984). Let us call this assumption about scientific explanations the "truth assumption." However, the truth assumption is inconsistent with the claim that models explain events or regularities. This is because models are idealized (i.e., literally false). Scientists routinely assert that their models explain. Clearly, something must give.

One can argue that the history of science provides evidence that the truth assumption is false. Consider a famous example (Kitcher, 1989, 453–454). According to lore, Galileo was asked the following why-question, "Why is the maximal range of a gun on a flat plane 45° rather than some other angle?" The answer can be provided if we make the following assumptions: the gun is a projector, the cannonball a point particle, the ambient atmosphere is a vacuum, and

---

Mangel, 1997, 34–36). This is the relationship between (1.2) and (1.5). When models are non-nested, then neither is a special case of the other. It is also important to note that model selection tools like the likelihood ratio test applies only to nested models, unlike the Akaike Inforation Criterion (Forster and Sober, 1994; Hilborn and Mangel, 1997; Burnham and Anderson, 2003).

[21] Robustness analysis is an important tool in modeling – for further discussion see Weisberg (2006); Weisberg and Reisman (2008); Wimsatt (2007); Odenbaugh (2011b); Odenbaugh and Alexandrova (2011).

the plane is an ideal Euclidean plane. Likewise, we assume the only force acting on the particle is gravity. Given these assumptions, it can be demonstrated for various angles of projection, that the maximal range of the gun is found at an elevation of 45°. Of course, it is unlikely that the fusiliers' gun was at a maximum range at exactly 45°. There must be surface irregularities in the barrel, air resistance, cannonball asymmetries, and the ground of the plane surely is not Euclidean. Nonetheless, it seems the why-question has been answered. Thus, models can successfully explain events or regularities.

Unfortunately, this argument from the history of science is unpersuasive. First, it clearly begs the question at issue. We cannot simply *assume* Galileo's model answered the why-question. Second, with models that purportedly explain, one can claim all we have is a "potential explanation" (Hempel, 1965a, 249). A potential explanation has all the features of an actual scientific explanation save the truth of the explanans.[22] This response requires us to deny what scientists often claim, namely that their idealized models explain. With historical case studies, then, we are at a stalemate.

Why believe that the truth assumption is true? Carl Hempel writes,

> That in a sound explanation, the statements constituting the explanans have to satisfy some condition of factual correctness is obvious. But it might seem more appropriate to stipulate that the explanans has to be highly confirmed by all the relevant evidence available rather than that it should be true. This stipulation, however, leads to awkward consequences. Suppose that a certain phenomenon was explained at an earlier stage of science, by means of an explanans which was well supported by the evidence then at hand, but which has been highly disconfirmed by more recent empirical findings. In such a case, we would have to say that originally the explanatory account was a correct explanation, but that it ceased to be one later, when unfavorable evidence was discovered. This does not appear to accord with sound common usage, which directs us to say that on the basis of the limited initial evidence, the truth of the explanans, and thus the soundness of the explanation, had been quite probable, but that the explanans is not true, and hence that the account in question is not – and never has been – a correct explanation (Hempel, 1965a, 248–249)

Hempel thus argues that, if we reject the truth assumption, then we have to say that an explanation can be correct at one time but not at a later one. But, this is inconsistent with "sound common usage." Therefore, we should accept the truth assumption. But it is not clear why ordinary language should dictate the

---

[22] In other words, potential explanations are how-possibly explanations. With a how-possibly explanation, the explanans do not contradict any know facts and, if true, would explain the explanandum. Idealizations of course do contradict known facts and thus cannot be part of a how-possibly explanation.

truth of the truth assumption. Suppose we did a survey and common usage went the other way – should we then reject the truth assumption?

A second argument for the truth assumption is offered by Wesley Salmon (1984). He writes,

> If one maintains that atoms and molecules are real entities, then it may be plausible to claim that they enable us to explain the behavior of gases. Pressure, for example, is explained on the basis of collisions of small material particles (which obey Newton's laws) within the walls of the container. If, however, atoms and molecules are mere fictions, it does not seem reasonable to suppose that we can explain the behavior of a gas by saying that it acts as if it were composed of small particles. (Salmon, 1984, 6)

Salmon suggests that, if something causally explains some phenomena, then the cause must exist. But, models introduce nonexisting properties and objects. Thus, models cannot provide causal explanations. As Nancy Cartwright (1983) has argued, "inference to the best cause" presupposes the existence of that which is causally responsible for the effect. Similarly, if we causally explain some phenomena by some theory or model, then the objects described must exist. However, the fact that a theory or model describes only objects or properties that exist is insufficient for the theory or model's truth. For example, presumably metapopulations really have probabilities of colonization and extinction. But, they are not actually constant in value. We are left with no cogent argument for or against the truth assumption.

Let's think about the truth assumption more carefully. Consider the following question and answer,

> "What time did Stacy arrive?"
> "She arrived at 5 pm."

It is extremely unlikely that she arrived *exactly* at 5 pm because she probably arrived before or after 5 pm. Thus, the answer is false although there is a sense in which it is approximately true. First, following Peter Smith (1998), "*p*" is approximately true if, and only if, approximately *p*. Thus, "She arrived at 5 pm" is approximately true if, and only if, approximately she arrived at 5 pm. Second, the answer is approximately true because of pragmatic features associated with the utterance or inscription even if false in terms of its semantic content alone. Linguist Peter Lasersohn (1999) has called these features "pragmatic halos." If we have an expression that denotes a set of objects, a pragmatic halo extends the set contextually. Thus, conversational context extends Stacy's arrival time around 5 pm and not exactly at it. Third, in ordinary conversations, we convey these pragmatic halos implicitly. However, in the sciences, we can explicitly create a pragmatic halo through statistics by providing best estimates and confidence intervals on our parameters. Remember our model (1.2) $dP/dt = cP(1 - P) - eP$. Suppose we derived the best fitting parameter estimates c_{est} and e_{est} along with confidence intervals, then our model would be of the form,

$$\frac{dP}{dt} = (c_{est} \pm x_1)P(1 - P) - (e_{est} \pm x_2)P \qquad (1.6)$$

Our model describes a subset of curves within a larger family. Hence, model (1.2) is approximately true if, and only if, approximately (1.2). But, approximately (1.2) is just (1.6) contextually specified (or something very much like it).[23] Fourth, we can explicitly create halos around our models, but we can also do so implicitly. Thus, I would revise the truth assumption – models can explain only if they are approximately true. Or, as I prefer to say, if they are "true enough" (Elgin, 2004, 2017; Teller, 2012).

1.5 Conclusion

In this section, I have argued that models are representations that idealize and abstract. Theories are clusters or families of models. And, we have seen how models can be explanatory through a notion of approximate truth. In the next section, we turn from what models are to how they are evaluated.

2 Testing Ecological Models

In this section, I start with a general methodology for how models are tested. From there, we dig in a bit on the topic of goodness-of-fit. We then consider an extended example of how ecological models are tested. Lastly, we see how models can be successful even when they are idealized and inaccurate.

2.1 A Methodology

I have suggested models are representations that idealize and abstract. We have also seen one common way ecologists talk about them as sets of assumptions. When we test models, we test those assumptions directly with observations or indirectly through their predictions. Of course, a model's predictions can be correct even when its assumptions are false. It is a fallible but reliable way of assessing them.

---

[23] Another way we see how pragmatic halos work, is to construe models as set-theoretic predicates (Suppes, 1957). For all x, x is a Levins's metapopulation if, and only if, (1) $x = < T, P, c, e >$, (2) T is an interval of times, (3) P is the proportion of patches occupied, (4) c and e are the probability of local colonization, (5) the probability of local extinction respectively, and (6) for all $t \in T$, then $dP/dt = cP(1 - P) - eP$. Pragmatic halos extend the extension of this predicate.

In considering the testing and confirmation of models in ecology, it is useful to consult the work of Elisabeth Lloyd (1994). Lloyd argues that the confirmation of models consists in three factors: (1) fit between model and data, (2) independent testing of aspects of the model, and (3) variety of evidence (Lloyd, 1994, 145). Let us examine each of Lloyd's features.

Lloyd writes,

> The most obvious way to support a claim of the form 'this natural system is described by the model' is to demonstrate the simple matching of some part of the model with some part of the natural system being described. (Lloyd, 1994, 146)

Lloyd has in mind here the confirmation of a model by confirming its predictions. Generally, a prediction of a model will be some statement concerning the dependent variables of the model (and possibly parameter values) (Lloyd, 1994, 147). The model, given initial conditions of the variables and auxiliary hypotheses concerning values of the parameters, will imply this statement or more generally will confer a likelihood on it. The likelihood is the probability of the prediction given the model (Hilborn and Mangel, 1997; Royall, 1997).

In order to test a model's predictions, we make a variety of observations that we record as data (Suppes, 1966; Bogen and Woodward, 1988). Hence, we can also misdescribe the data itself.[24] We check those predictions against our data. Given our notion of prediction, we might simplistically think of a model as confirmed if the model predicts the data. Likewise, the model is disconfirmed if it does not predict them (provided those data are relevant to testing the model). This, as you might suspect, is too simplistic. First, it is rare that we simply observe an event or state of affairs predicted by the model. Rather, we make observations and record data always with measurement error. Second, we infer the properties of a population, community, or ecosystem on the basis of a sample, which can be another source of error.[25] Third, if the data predicted are used to parameterize the model, then successful prediction might not count as confirmation.[26] Given these qualifications, it is most usual

---

[24] For example, Paine (1988) has argued that descriptions of food webs are extremely simplified and, hence, testing models on the basis of food web data is highly misleading at best. In virtue of moving from observations to data, much of the ecological system studied is ignored and, according to Paine, it should not be.

[25] For an interesting short discussion as to whether food webs are observable, see Sober (1985, 15–16). We observe humans eating food. They are a part of a food web. If by observing x, which is a part of y, we observe y, then we observe the food web. But, food webs have properties like *connnectance* and *linkage density* that are not observable. Maybe the right answer is that some properties of food webs are observable and others not.

[26] Some philosophers and scientists think that, if you use data in formulating a model, the data cannot confirm the model or not nearly as much as predicting the data without using it to

and appropriate to evaluate a model's predictions via its *goodness-of-fit* to the phenomena.

> Fit can be evaluated by determining the fit of one curve (the model trajectory or coexistence conditions) to another (taken from the natural system); ordinary statistical techniques of evaluating curve fitting are used. (Lloyd, 1994, 147)

In the previous section, I was critical of the similarity view. According to that view, our mathematical equations describe a mathematical structure that is similar to an ecological system in certain respects to certain degrees. They are similar because they share properties, and I argued that they don't share the relevant ones. According to a deflationary view, mathematical models have adjustable parameters that specify a family of curves (Sober, 2002). We find the best estimates of those parameters and compare the fitted model to the system of interest. Additionally, we already have the relevant measures of similarity between model prediction and ecological system available. We find them in statistics and model selection. We do not need a measure of similarity *de novo* because statisticians have already developed them.

An increasingly popular approach to model fitting goes like this (Brown and Rothery, 1993; Hilborn and Mangel, 1997; Burnham and Anderson, 2003). First, we must choose the functional form of our model, which provides us with a family of curves.[27] Often the baseline or simplest form is a linear model,

$$y = ax + b + \epsilon \tag{2.1}$$

where y is a dependent variable, x is an independent variable, a is a slope, b is the y-intercept, and ϵ is an error term. We can add complexity to our model by including higher-degree polynomial expressions like x^2 or additional independent variables $x_1, x_2, \ldots, x_n$. Second, we find the best estimate of our parameters that selects a curve out of the family and bests fits the data. For example, we must estimate the values of our parameters a and b in (2.1). These are the values such that our model's predictions regarding the dependent variables fit the data best. We might use least sum of squares where the differences between model predictions and data are minimized. Or, we might use a maximum likelihood method of parameter estimation.

---

parameterize your model. That is, prediction provides more evidence than accommodation. For a discussion in the context of model testing, see Hitchcock and Sober (2004). A common strategy is to parameterize or initialize your model with part of a time series of data and to test the model with a different part of that time series.

[27] The functional form of the equation could be linear, quadratic, power, polynomial, exponential, logarithmic, and so on.

Technical Discussion

The simplest approach for determining a model's goodness-of-fit to data is the sum of squares. Suppose we have a system with independent variables $x_1, x_2, \ldots, x_n$ and dependent variables $y_1, y_2, \ldots, y_n$ (see Hilborn and Mangel, 1997). Our dummy model is,

$$y_i = a + bx_i + cx_i^2 + \epsilon_i \qquad (2.2)$$

where ϵ_i is a term for uncertainty and where a, b, and c are parameters. Suppose that there is no observational error ($\epsilon_i = 0$). For each value of the parameters and value of the independent variable, we determine a value of the dependent variables. So, for a given value of x and for estimated values of the parameters a_{est}, b_{est}, and c_{est}, we predict the value of y to be,

$$y_{est,i} = a_{est,i} + b_{est,i}x_i + c_{est,i}x_i^2 \qquad (2.3)$$

The next step in our sums of squares analysis is to measure the deviation between the i-th predicted and observed values by the square $\left(y_{pre,i} - y_{obs,i}\right)^2$. Next, we sum the deviations over all of the data points, and obtain,

$$\vartheta(a_{est}, b_{est}, c_{est}) = \sum_{i=1}^{n}(y_{pre,i} - y_{obs,i})^2 \qquad (2.4)$$

$$= \sum_{i=1}^{n}(a_{est,i} + b_{est,i}x_i + c_{est,i}x_i^2 - y_{obs,i})^2 \qquad (2.5)$$

Now we have a measure of fit between the model and the data if the parameters are a_{est}, b_{est}, and c_{est}. Finally, we find the partial derivatives of (2.5) with respect to a_{est}, b_{est}, and c_{est} and set the equations to zero; that is, $\frac{\partial\vartheta(a_{est},b_{est},c_{est})}{\partial a_{est}} = \frac{\partial\vartheta(a_{est},b_{est},c_{est})}{\partial b_{est}} = \frac{\partial\vartheta(a_{est},b_{est},c_{est})}{\partial c_{est}} = 0$. This results in three equations with three unknowns that can be solved in order to give us the parameter values a^*, b^*, and c^*, which minimizes the sum of squares or makes it as small as possible, that is, $\vartheta_{min} = \vartheta(a^*, b^*, c^*)$.

There are limitations to sums of squares as a statistic for determining the fit of a model or models to data. First, the independent variables must be genuinely independent. Second, there is a deterministic relationship assumed between the independent variables and dependent variables. There are other more complex goodness-of-fit procedures that can be used (see Brown and Rothery, 1993; Hilborn and Mangel, 1997). Nevertheless, once we have completed these two steps, we can compare different models and their fit to the data.

Model fitting might seem innocuous, but there are big philosophical issues at stake (Taper and Lele, 2010). Suppose one is only concerned with a model's fit to the data. One should then always choose a more complex model since it provides a better fit to the data than any less complex model. For example, a higher-degree polynomial model will always have a smaller sum-of-squares score than a linear model. However, more complex models risk overfitting the data. Overfitting occurs when a model is "too sensitive" to the peculiarities of a data set, which are not present in other data sets, and thus are less accurate of future data (Hitchcock and Sober, 2004, 11). Additionally, the simplicity of a model has often been thought of as an aesthetic feature and not an epistemic one. A recent popular framework seems to allay these worries – the Akaike Information Criterion (AIC) (Forster and Sober, 1994; Forster, 2000, 2002; Sober, 2002). The AIC for a model M_i with p_i parameters given data Y is (Hilborn and Mangel, 1997, 159),

$$A_i = \mathbf{L}(Y|M_i) + 2p_i$$

AIC adds 2 for each of the adjustable parameter p_i to the negative log-likelihood $\mathbf{L}(Y|M_i)$ and thus penalizes overfitting.[28] Thus, AIC combines goodness of fit, the negative log-likelihood, and the simplicity, the penalization of additional parameters, of models into a common currency.

In ecology, one concern regarding AIC is that it requires a "uniformity of nature" assumption (Moll et al., 2016). This assumption claims that the data used to estimate the parameters of the model and the data about which predictions are tested come from the same probability distribution.[29] Moll et. al. argue that ecological systems like the proposed wolves–elk–vegetation trophic cascade in Yellowstone National Park are very complex insofar as there is a large number of independent variables for each dependent variable, the variables are hierarchically structured, they interact with one another, and they exhibit nonstationary distributions. Thus, critics claim complex systems like populations, communities, and ecosystems are unlikely to satisfy the uniformity of nature assumption.[30]

---

[28] AIC has the other advantage that it applies to non-nested models, unlike the likelihood ratio test.

[29] For example, suppose our parameter of interest is X that is a continuous random variable that has a probability density function $f(x)$. The probability of falling into an interval $[a, b]$ is given by,

$$\Pr[a \leq X \leq b] = \int_a^b f(x)dx$$

The uniformity of nature is the assumption that $f(x)$ doesn't change over time.

[30] It is worth noting that the uniformity of nature assumption is common across model selection criteria and is not unique to AIC.

Lloyd proposes that another factor in the confirmation of models is the testing of independent assumptions of a model.

> Numerous assumptions are made in the construction of any model. These include assumptions about which factors influence the changes in the system, what the ranges for the parameters are, and what the mathematical form of the laws is. Many of these assumptions have potential empirical content ... [W]hen empirical claims are then made about this model, the assumptions may have empirical significance. (Lloyd, 1994, 147)

Assumptions of models concern features of the following form: the choice of variables, the choice of parameters (constant values or random variables), and the choice of the mathematical form of the laws of succession and coexistence (continuous or discrete, deterministic or stochastic). In the construction of models, modelers make particular assumptions about the dynamical systems that they are working with. However, these assumptions all come in either one of these forms. What makes an assumption different from a prediction? Assumptions are of a general form, whereas predictions are typically specific values of the dependent variables implied by the assumptions and auxiliary hypotheses.

What makes assumptions of a model independent? One might construe independent assumptions as assumptions that are independent in the traditional statistical sense (Lloyd, 1994, 149). That is, the events or random variables those assumptions describe must be independent.

TECHNICAL DISCUSSION

Assumptions of models may describe events or random variables. Events E_1 and E_2 are independent if, and only if, $\Pr(E_2, E_1) = \Pr(E_2) \times \Pr(E_1)$. If we are talking about random variables X and Y, with cumulative distribution functions $F_X(x)$ and $F_Y(y)$ and probability densities $f_X(x)$ and $f_Y(y)$, then X and Y are independent if, and only if, the combined random variable (X, Y) has a joint cumulative distribution function $F_{X,Y}(x, y) = F_X(x)F_Y(y)$ (or a joint density $f_{X,Y}(x, y) = f_X(x)f_Y(y)$). Thus, on a frequentist approach, model assumptions are independent if the events or random variables they describe are independent.

However, for many of these propositions, their respective probabilities are lost on us. From a frequentist interpretation, we must determine the frequency of an event or of a random variable taking some set of values in an experimental setup (Hacking, 2016, ch. 2). But, for many model assumptions, there will be no such experimental setup by which to determine their respective probability. On a Bayesian interpretation, one could regard model assumption probabilities as degrees of belief (Howson and Urbach, 2006). But, this inherits all of the worries (and benefits) regarding subjectivism about probability. Thus, insofar as our model assumptions do not concern experimental setups for events or random variables, we minimally say independent assumptions as ones with *logically independent* assumptions; neither assumption entails the other.

Why would it matter whether or not two or more assumptions of a model are independent of one another? This question leads us to Lloyd's last confirmational feature – variety of evidence. Variety comes in, well, varieties, and so we will consider two types. First, ideally we would like our model's goodness-of-fit profile to be determined through a variety of data sets. For example, we might test our model against different experimental populations of the same species. Or we might test it against different species of the same genus, among others. A model that fits a greater number of data types is more confirmed than one that fits fewer *ceteris paribus*. Second, a model that has evidence for more independent assumptions is more confirmed than one that has evidence for fewer independent assumptions. One justification for this claim is that diverse data will eliminate many of the most plausible competitors, whereas narrow data leave many plausible competitors still standing (Horwich, 2016).[31]

Models are tested through their goodness-of-fit to data and their independent assumptions where variety in both is crucial. Let's now turn to an example.

2.2 Blowflies and Delay Differential Equations

Ecologist A. J. Nicholson (1954, 1957) did important laboratory experiments on time delays in population cycles of the sheep blowfly *Lucilia cuprina*. Nicholson noticed population cycles in his blowflies, and he wanted to know what brought them about. He hypothesized that the cycles were due to a time delay in the response of fecundity and mortality to the abundance of adults. One such confirming experiment involved Nicholson restricting blowfly larvae to 50 g of liver each day, even though the adults received unlimited food. As the result, the adult population increased to a maximum of about 4,000 individuals and then crashed to a minimum of 0, at which time all individuals were eggs or larvae, and the cycle would start again. The period of the cycle was between thirty and forty days. If the adults were numerous, then lots and lots of eggs would be laid, which caused extreme larval competition for the 50 g of liver. During the adult population peak, none of the eggs survived to

---

[31] Here is a sketch of Paul Horwich's (2016, 118–119) justification of variety. Suppose we have a set of mutually exclusive and exhaustive set of hypotheses $H_1, H_2, \ldots, H_n$. Also suppose we have a diverse set of evidence E_D and a narrow set E_N, both of which are entailed by H_1. For any $Pr(H_i)$, $Pr(E_D|H_i) > Pr(E_N|H_i)$. Thus, the confirmatory boost provided by E_D will be greater than E_N; that is, $Pr(H_1|E_D) > Pr(H_1|E_D)$. For an alternative, see Earman (1992, 77–79).

pupate. Thus, the large adult populations did not give rise to surviving offspring and the adult population began to shrink. After about a month with lessened competition, offspring would survive and, thus, the adult population would increase.[32]

Robert May (1973a) examined Nicholson's experimental work on blowflies and fit a model to the time series. May modeled the blowfly population dynamics with the delayed logistic equation

$$\frac{dN(t)}{dt} = rN(t) \left(1 - \frac{N(t-T)}{K} \right) \tag{2.6}$$

where r is the intrinsic rate of growth, $N(t)$ is the population size at time t, K is the carrying capacity, and T is the time lag. If $T = 0$, then we have the logistic equation,

$$\frac{dN(t)}{dt} = rN(t) \left(1 - \frac{N(t)}{K} \right) \tag{2.7}$$

with a stable equilibrium when $N = K$. May wanted to know the answer to two questions: (1) Can the delayed logistic generate stable limit cycles?; and (2) Can the delayed logistic provide a good fit to the blowfly data?

Let's consider the first question. If $rT = \tau < 1/2\pi$, then there remains a stable equilibrium point at K. If $rT = \tau > 1/2\pi$, then there are stable limit cycles and the oscillations increase as rT increases. When $N/K = 1$ and $\tau = 0$, then there is a stable equilibrium. However, as τ increases ($\tau_1 = 1.6$, $\tau_2 = 1.75$, $\tau_3 = 2.0$, $\tau_4 = 2.5$), then N/K begins to increase and then decrease more violently in the cycle. Thus, May answered his first question – the delayed logistic can exhibit stable limit cycles.

In order for May to determine the model's goodness of fit of the delayed logistic model to Nicholson's data, we must put the delay differential equation into canonical form by letting $\tau = rT$,

$$\frac{d\hat{x}}{dt} = x(\hat{t}) \left(1 - x(\hat{t} - \tau) \right) \tag{2.8}$$

where $x = N/K$ and $\hat{t} = rt$ so that there would be one dimensionless parameter rT to estimate.[33] May estimated from the observed time series that $N_{max}/N_{min} = 84$ from the fourth (largest) cycle. So, from a table of analytic

---

[32] Nicholson demonstrated this by removing the time delay (1957, 158). He did this by reducing the adult food intake to 1 g per day, which reduced the number of eggs produced. As the result, the population cycling largely disappeared.

[33] A parameter is dimensionless if the parameter has no units. In this case, r has inverse units of time and T has units of time. Thus, the units cancel, and the parameter $rT = \tau$ is dimensionless. Dimensionless parameters are important in stability analyses since the stability of a system should not depend on the units of the parameters.

properties of the delayed logistic (May, 1973a, 97), if $N_{max}/N_{min} = 84$, then $rT \approx 2.1$ and thus the period is $4.54T$. May knew that there are six full cycles in the data set and the six cycles occur in 232 days, so we have the following equality, which we can solve for T, $6 \times (4.54T) = 232$ days.

$$27.24T = 232 \text{ days}$$
$$T = 232 \text{ days}/27.24$$
$$= 8.46 \text{ days}$$

So, the predicted time lag between eggs and adults is approximately 8.5 days and the observed delay is 14 days.

May writes,

> In view of the crudity of the model, the agreement is surprisingly good. It suggests that stable limit cycles generated by the time-delayed regulatory mechanism are indeed the basic feature in the dynamics of these populations. (May, 1973a, 102–103).[34]

There are some important questions we can raise concerning how well May's delayed logistic model fits Nicholson's data. Gurney et al. (1980), for example, raise two objections. First, the difference between the predicted delay of 8.5 days is not terribly close to the observed delay of 14 days. Can another model do better? Second, the time-delay logistic model cannot predict the two bursts (i.e., "double hump") of reproductive activity per adult population cycle. They write, "It is thus structurally incapable of explaining the results obtained by Nicholson in the larval-limited regime" (Gurney et al., 1980, 18). Can another model predict it?

Gurney et al. propose their own more realistic birth–death model,

$$\frac{dN}{dt} = R - D = R - \delta N$$

where R is the recruitment rate and D is the mortality rate.[35] They assume that the per capita death rate has a time- and density-independent value δ. They additionally assume the rate at which eggs are produced depends only on the current size of the adult population, eggs develop into sexually mature adults in T_D time units, and the probability of an egg developing into a viable adult only depends on the number of competitors of the same age. Thus, rate of recruitment at t is a function of the adult population size at $t - T_D$.

---

[34] We can determine the fit of a model to data following the procedures described earlier. However, it is arguable that determining whether the fit is *good* requires comparing different models. If correct, then goodness of fit is a *comparative* notion. For further discussion of these issues by philosophers, ecologists, and statisticians, see the essays in Taper and Lele (2010).

[35] Recruitment occurs when juvenile organisms enter into the adult age or size class.

$$R = RN(t - T_D)$$

This incorporates more realistic biological details than May's model. Therefore,

$$\frac{dN}{dt} = RN(t - T_D) - \delta N(t) \tag{2.9}$$

Examining Nicholson's data, they argued any plausible functional form for $RN(t - T_D)$ must go to zero as N becomes either very large or very small. Also, most recruitment curves display a single maximum. They chose the following simple function that has all of these features,

$$RN(t - T_D) = PN(t - T_D) \exp\left(\frac{-N(t - T_D)}{N_0}\right)$$

where P is the maximum per capita fecundity and N_0 is the adult density at which maximum reproductive success occurs. Our final model is,

$$\frac{dN(t)}{dt} = PN(t - T_D) \exp\left(\frac{-N(t - T_D)}{N_0}\right) - \delta N(t) \tag{2.10}$$

This model has a single equilibrium,

$$N^* = N_D \ln(P/\delta)$$

The local stability of this equilibrium and the qualitative behavior around it are determined by PT_D and δT_D.

If the parameters of our model are in the stable/underdamped region, we expect stochasticity to produce quasi-cyclic population fluctuations. Those fluctuations will be coherent, meaning that the period is close to the deterministic natural period with noise. We can describe the return of the undamped population to its deterministic natural period by a "coherence number" n_c. This is the number of cycles over which the amplitude is reduced by a factor e. This quasi-cyclic behavior in the stable/underdamped region is characterized by the normalized cycle period (T/T_D) and n_c on the $PT_D/\delta T_D$ plane. If the parameters are in the locally unstable region, the solutions can be obtained by numerical integration, which can take the form of a limit cycle. Analysis reveals that the solution is structured around a dominant period T. The solutions then are characterized in this region by the ratio of the dominant period to the delay (T/T_D) and the ratio of maximum to minimum population (N_{max}/N_{min}). We can plot this in the appropriate part of the $PT_D/\delta T_D$ plane too. We can test whether quasi-cycles or limit cycles are occurring by finding the correct values of these quantities. Given that $T_D = 14.8 \pm 0.4$, these are,

$$2.5 < (T/T_D) < 2.7; 29 < (N_{max}/N_{min}) < 53; 2 < n_c < 5$$

The best-fitting values for PT_D and δT_D on the stable limit cycle hypothesis are,

$$PT_D = 150 \pm 70; \delta T_D = 2.9 \pm 0.5$$

and for the quasi-cycle hypothesis,

$$PT_D = 23.5 \pm 4.5; \delta T_D = 3.0 \pm 0.7$$

Although δT_D values are consistent with each hypothesis, we can test them with their PT_D values. From data, Gurney et al. (1980) determine the independent estimate to be,

$$PT_D = 130 \pm 30$$

which thus confirms the limit cycle hypothesis and disconfirms the quasi-cyclic hypothesis.

Additionally, Gurney et al. (1980) simulated model (2.10) with parameter values taken from Nicholson's data, which produced the two discrete "bursts" or humps that May could not produce with model (2.6). In the model, the distribution of adult-state recruitment within each cycle depends on how low the adult numbers fall. If the minimum adult number $N_{min} > N_0$, then recruitment into the adult shows a single peak. If $N_{min} < N_0$, then the single peak begins to split. Finally, if $N_{min} << N_0$, then there are two or more peaks per cycle. This simulations corresponds well to when adult food is limited (single double-hump peak per cycle) and when larvae are food limited (clear generations occurring at uneven time periods).[36]

The work on Nicholson's blowflies by May and Gurney et al. show how ecological models are tested. May proposed the delayed logistic, and he found a good but coarse-grained fit. There was also the double hump that his model could not generate. Gurney et al. added more realistic assumptions about the recruitment function and adult mortality and produced a better fit, eliminating a rival hypothesis. Additionally, their model was able to generate the elusive double hump that May could not. Ecologists test models through evaluating their assumptions and through their fit to data.[37]

---

[36] For more recent work on modeling blowflies with increasingly sophisticated techniques for model fitting, see Kendall et al. (1999). For a helpful discussion of these episodes, see Renshaw (1993, 116–127).

[37] One might wonder can May's and Gurney et al.'s modeling of laboratory populations tell us about populations in the field? This is an excellent question, and I have argued that, if models cannot provide a good fit in so-called bottle experiments, they are unlikely to do so in the wild (Odenbaugh, 2006b). Thus, it can be a good start to determine which models are worth pursuit. For a different view, see Currie (2018).

2.3 Coda: Inaccurate and False Models Can Be Successful

Our models are often inaccurate or false. Also there is often a "backlog" of models in ecology that have yet to be tested. Some have argued that, as a result, ecology is a "sick science." Daniel Simberloff writes,

> Ecology is awash in all manner of untested (and often untestable) models, most claiming to be heuristic, many simple elaborations of earlier untested models. Entire journals are devoted to such work, and are as remote from biological reality as are faith-healers. (Simberloff, 1981, 3)

The previous section's examples should make one pause when one hears this work is like that of "faith-healers."[38] However, models when untestable, untested, or predictively inaccurate can still be successful. In Section 1, we characterized idealizations as false assumptions that are *useful*. They can be useful, for example, when they can be used to explore possibilities and refine concepts. Let me mention two examples to make the point.

2.3.1 Chaos and Population Regulation

One of the very first demonstrations of chaotic behavior comes from Robert May (May, 1973a, 1974, 1975, 1976; May and Oster, 1976). Using simple models, he showed ecological systems could exhibit chaotic behavior. Informally, a dynamical system exhibits chaos if it is extremely sensitive to initial conditions and is aperiodic. A dynamical system exhibits sensitive dependence on initial conditions if, and only if, for arbitrarily close initial conditions the subsequent behaviors of the dynamical system diverge exponentially (this is sometimes called the "butterfly effect"). A system is aperiodic if its trajectories never repeat their same sequences. Chaotic systems are, however, bound within a neighborhood called a "strange attractor."

Here is an example May used,

$$N_{t+1} = N_t + RN_t \left(1 - \frac{N}{K}\right) \tag{2.11}$$

---

[38] For similar but less spirited worries see Peters (1991), Shrader-Frechette and McCoy (1993) and for responses see Cooper (2003), and Odenbaugh (2005). There is a deeper issue beneath Simberloff's worry that I think is very important. The issue is one of the social structure of ecology. How many ecologists should be theoreticians, experimentalists, or field ecologists? Clearly, if there are too many theoreticians, then too little data is collected and too few models are tested. If there are too few, then we amass data, but have difficulty in formulating hypotheses and testing them. I am assuming that that these are distinct groups, but, obviously, some ecologists find themselves in more than one cohort. In my estimation, it is worth considering these allocation problems using the strategies outlined by Kitcher (1995) and Strevens (2003).

N_t represents population size or density at time t, K is the carrying capacity of the population, and R is the net rate of reproduction. In our model, if $R \leq 2.57$, then for various values of R the model exhibits a stable equilibrium and stable cycles. However, if $R > 2.57$, then the model exhibits chaos.

Here is another famous example of May's. Suppose there is a seasonally breeding population of insects whose generations do not overlap. Modeling their population dynamics with a discrete law of succession relates the population at $t + 1$ to that of t, or $x_{t+1} = F(x_t)$. The simplest useful expression is a nonlinear difference equation called the "logistic map,"

$$x_{t+1} = ax_{t+1}(1 - x_{t+1}) \tag{2.12}$$

In this model, x_t is normalized; it is scaled by dividing population size by the maximum value, and so $0 < x_t < 1$. If $a < 1$, then the population will decrease to extinction; however, if $a > 1$ the dynamics of our model can become very complex. At $a = 2$, it is stable as a point attractor. At $a = 3.3$, it displays periodic behavior as limit-cycle attractors, with the number of available attractors doubling as a increases. Lastly, chaotic behavior occurs in a strange attractor at $a = 3.56995$.

TECHNICAL DISCUSSION

It is very hard to discover whether a given ecological system is chaotic (Hastings et al., 1993; Cushing et al., 2002). Ecologists analyze time series to assess whether such systems are density dependent and sensitive to initial conditions. One can quantify this sensitivity through a measure called a Lyapunov exponent. Suppose we have an initial population of size N_0 and consider a nearby population of size $N_0 + \Delta_0$, where Δ_0 is very, very small. After n time steps we can examine the sizes of two respective populations, the first starting at N_0 and the second at $N_0 + \Delta_0$. Let $|\Delta_n|$ be the absolute value of their difference. We do this again and again and then fit this data to an exponential model with an elapsed time of n as the independent variable,

$$|\Delta_n| = |\Delta_0| e^{n\lambda}$$

This resulting model has one parameter, the Lyapunov exponent λ (Case, 2000, 116–117). If $\lambda < 0$, then populations are converging over time; if $\lambda > 0$, then the difference between populations at the initial times is growing exponentially over time, which is indicative of chaos.

One extremely important implication of May's work on chaos concerns debates over population regulation. Population ecologists have been arguing for quite some time over whether or not populations are regulated. Many populations persist through time and their abundance has a mean value with a moderate variance. The biotic school was developed by Howard and Fiske (1911), Nicholson and Bailey (1935), and Smith (1935), and they argued that populations were regulated by density dependent factors. A density dependent factor is one whereby the rate of population change is a function of the population size itself. The climatic school pioneered by Bodenheimer (1928) and Andrewartha (1954) argued that populations are driven by changes in the abiotic environment like weather, and thus fluctuate greatly. These factors are density independent, and they kept populations low enough that density dependence could not occur.

As one example of the debate, Davidson and Andrewartha (1948) published a study on *Thrips imaginis*, a plant-sucking insect, which live in roses grown in Adelaide, Australia. They counted the number of *Thrips* in a sample of twenty roses every week for fourteen years. Each year, the number of *Thrips* reached a peak value toward the end of the month of November. Davidson and Andrewartha attempted to determine the effect of weather on population abundance by finding correlations with meteorological factors using multiple regression. In Adelaide, the winters are cool and rainy and the summers are hot and dry. So, *Thrips* do better in the summer and only in the winter months do they reach the peak and then quickly decline. The independent meteorological variables of the model accounted for 78 percent of the variance. Davidson and Andrewartha concluded that *Thrips* ate little of the food available and the variation in abundances were accurately captured by variation in abiotic factors.

Smith (1961) replied that regression could not detect density dependence in *Thrips*. First, if one is to detect detect density-dependence one must examine relative not simply absolute population abundance (Smith, 1961, 403). By estimating the rate of population change from average logarithms $\overline{\Delta \log N}$ and population size $\log N$, he found that they were negatively correlated immediately preceding the spring peak (Smith, 1961, 404). This is evidence of density dependence. Second, Smith noted that, if only density independence were occurring, then the variance of the logarithm of average population size $\log \overline{N}$ should increase with population size $\log N$. However, the variance decreases as population increases in the months of October, November, and December (Smith, 1961, 406).

May argued that a key assumption of this complex debate was flawed. Equations (2.11) and (2.12) show that density dependent factors can produce the very same type of "random" population dynamics that appear to be due to density independent factors such as weather.

> These studies of the Logistic Map revolutionized ecologists' understanding
> of the fluctuations of animal populations …With the insights of the Logistic
> Map, it was clear that the Nicholson-Birch controversy was misconceived.
> Both parties missed the point: population-density effects can, if sufficiently
> strong …, look identical to the effect of external disturbances. The problem is
> not to decide whether populations are regulated by density-dependent effects
> (and therefore steady) or whether they are governed by external noise (and
> therefore fluctuate). It's not a question of either/or. Rather, when ecologists
> observe a fluctuating population, they have to find out whether the fluc-
> tuations are caused by external environmental events (for example, erratic
> changes in temperature or rainfall), or by its own inherent chaotic dynam-
> ics, as expressed in the underlying deterministic equation that governs the
> population's development. (May, 2002, 223)

Ecologists attempt to demonstrate chaos in ecological systems where they suspect it might be found. However, even if those models are ultimately untestable,
May's models provided us the means to explore ecological possibilities and
challenged fundamental assumptions in important ecological debates.[39] As
May writes,

> It seemed to some that only one of these two pictures could be right. But, as
> often happens in science when two opposing views of a problem both seem to
> be partially right yet ultimately irreconcilable, many of the protagonists were
> looking at the problem in too narrow a way. The problem, it turned out, can
> be understood much more simply in terms of a different way of thinking,
> of a different paradigm. The virtue of the Logistic Map was that it gave a
> clear and easy-to-grasp idea of this new and extraordinarily productive way
> of thinking, as I was soon to find out. (May, 2002, 217)

Building models can make our thinking more rigorous and force us to make
explicit our assumptions. In doing so, unknown empirical possibilities are
opened up for investigation and testing.[40]

Sometimes these new methods and models result in unexpected empirical successes. Recently, an "equation-free" approach to modeling nonlinear
dynamics has appeared in the context of fisheries (Sugihara et al., 2012; DeAngelis and Yurek, 2015; Ye et al., 2015).[41] The traditional approach we have
been exploring in this Element uses parameterized models to represent causal

---

[39] For more discussion on the debates over population regulation, see Cooper (2003) and Odenbaugh (2006c), and for discussions of chaos in ecology, see Costantino et al. (1997), Hastings
et al. (1993), and Odenbaugh (2006b, 2011a). See Kellert (1994) and Smith (1998) for a
philosopher's analysis of mathematical chaos more generally.

[40] May's work demonstrated the possibility of chaos in ecological populations. His models by
themselves, of course, don't show that it actually occurs. And, even if it can be demonstrated in
the laboratory, which it has, this doesn't show it occurs frequently in the field.

[41] For an introduction, see Chang et al. (2017).

processes. For example, we identity variables and parameters of interest and specify the functional form of the equations using observed correlations. Thereafter, we find the best estimates of the parameters and make predictions. However, those can be "mirage correlations," which transiently appear through nonlinear dynamics (Ye et al., 2015, E1570). Empirical data modeling (EDM) takes a very different approach. Suppose we have a dynamical system that consists in three variables; namely, phytoplankton $x(t)$, zooplankton $y(t)$, and fish $z(t)$. The set of values for all of the variables is a phase space or manifold. Observations of the manifold are times series that are sequences of observations of a variable over time $\{x_t, x_{t-2}, \ldots, x_{t-n}\}$. For example, it would include fish abundances at just those times. If we had a time series for every variable, we can reconstruct the "attractor manifold." Taken's theorem astonishingly shows that information regarding the entire system can be reconstructed from a sufficiently long single variable time series. To do so, we take single variable time series, say $x(t)$ of our phytoplankton, and use lagged coordinate embedding $\{x_t, x_{t-1\tau}, x_{t-2\tau}, \ldots, x_{t-(E-1)\tau}\}$. E is the embedding dimension that is the dimension or number of time-delayed coordinates necessary to reconstruct the attractor, and τ is the lag. This can be plotted against population size $x(t)$ and an isomorphic phase portrait reconstructed. This new approach has already been used to study sockeye salmon in the Fraser River in British Columbia. And, it has been far more accurate than traditional Ricker recruitment models.[42] Ye et al. (2015) predicted a run of salmon between 4.5 and 9.1 million, whereas traditional models predicted between 6.9 and 20 million. The correct answer was about 8.8 million.

2.3.2 Food Webs and Stability Concepts

Our second example of how inaccurate and false models contribute to successful ecology comes from the work of May as well. During this same period, he was exploring the relationships between complexity and stability. Following Gardner and Ashby (1970), May constructed his model communities with m species by choosing the interaction coefficients a_{ij} in a $m \times m$ community matrix. a_{ij} represents the per capita effect of species j on species i. Adding

---

[42] William Ricker (1954) introduced what is naturally called the "Ricker recruitment model," which is a discrete population model that provides the expected number N_{t+1} of individuals in generation $t + 1$ as a function of the in previous generation t,

$$N_{t+1} = N_t e^{r\left(1-\frac{N_t}{k}\right)}$$

where r is an intrinsic growth rate and k the carrying capacity. Ricker models are extremely common in fisheries science.

Table 2.1 Types of species
Interactions. (May, 1973b, 639)

		a_{ij}		
		+	0	−
	+	++	+0	+−
a_{ji}	0	0+	00	0−
	−	−+	−0	−−

another individual of species j can have a negative, positive, or neutral effect on i.

There are five different categories of interaction between any pair of species commensalism (+0), amensalism (−0), mutualism or symbiosis (++), competition (−−), and predator–prey (+−). He assumed each species exhibited intraspecific competition so that $a_{ii} = -1$. He defined connectance C as the probability that two species interact, which is measured by $a_{ij} \neq 0$. Thus, $(1 - C)$ are those $a_{ij} = 0$. The a_{ij} were set at random. Thus, some were greater than, less than, or equal to zero. The intensity s of a_{ij} is a random variable with a mean of zero and variance of s^2.

TECHNICAL DISCUSSION

May constructs model communities with m species with densities $N_i(t)$ whose dynamics are governed by nonlinear first-order differential equations.

$$\frac{dN_i(t)}{dt} = F_i\left((N_1(t), N_2(t), \ldots, N_m(t)\right)$$

The possible equilibria N_i^* of the species occur when $dN_i(t)/dt = 0$. We can study the stability of an equilibrium community by examining the dynamics of,

$$N_i(t) = N_i^*(t) + x_i(t)$$

where $x_i(t)$ refers to the arbitrarily small perturbations to the equilibrium $N_i^*(t)$. If we provide a Taylor expansion of the n equations around the equilibrium, we have a set of n linear first-order differential equations,

$$\frac{dx_i}{dt} = \sum_{j=i}^{m} a_{ij} x_j(t) \tag{2.13}$$

where a_{ij} is the coefficient of interaction between species i and j, which is the effect of species j on i, as we have seen. We can describe the possible interactions among i and j by the signs of a_{ij}. We can also represent (2.13) as,

$$\frac{d\mathbf{x}}{dt} = \mathbf{A}x(t)$$

where $\mathbf{x}$ is the $m \times 1$ column vector of x_i and $\mathbf{A}$ is the $m \times m$ community matrix whose elements a_{ij} describe the effect of species j on species i near equilibrium or,

$$a_{ij} = \left(\frac{\partial F_i}{\partial N_j}\right)^*$$

An equilibrium point is locally stable just in case all of the eigenvalues of the community have negative real parts.

May then proved that a model community is "almost certainly" stable in the case of,

$$s\,(mC)^{\frac{1}{2}} < 1$$

This had a surprising consequence. Suppose $\Pr(m, s, C)$ is the probability that a model community is stable. Then $\Pr(m, s, C) \rightarrow 0$ as $m \rightarrow \infty$ and $\Pr(m, s, C) \rightarrow 1$ as $m \rightarrow 0$. Therefore, all else being equal, as the number of species in a community increases, then the probability that the community is stable decreases. This was so surprising because it was commonly thought that, as the number of species increases in a community, the more likely it would be stable. Communities exhibited a "balance of nature" (Egerton, 1973; Kricher, 2009). The system would more likely return to a steady state after a perturbation, especially with a large number of interconnected species. This model really fundamentally challenged this piece of folk ecology; one could not simply *assume* more complex communities are more stable.

May's model, with its assumptions, has been challenged (De Angelis, 1975; Lawlor, 1978; Mikkelson, 1997; McCann, 2005). For example, his communities can have predators with no prey and prey without predators, and can also have food web loops where species i feeds on j, j feeds on k, and k feeds on i, which are considered biologically unrealistic. Peter Yodzis (1981) for example argued a community matrix is a "truly local object" and he built "plausible" community matrices. For example, if i is prey and j is predator, then $a_{ij} < 0$ and $a_{ji} > 0$. However, the effect of i on j is much smaller than j on i. If i and j are competitors, then $a_{ij} < 0$ and $a_{ji} < 0$. But they need not be symmetrical in their magnitude.[43] Lastly, May assumed each species exhibits density

---

[43] Yodzis argues that they might be symmetrical when competing over resources but not over space.

dependence. But, not all species exhibit density dependence. If one construct models communities when these more realistic interaction coefficients, then stability is more likely in these communities than in May's.

Nevertheless, as ecologists questioned May's model, they came to recognize a whole suite of distinct concepts (Holling, 1973; Orians, 1975; Pimm, 1984). Stuart Pimm (1984) distinguished between *complexity*, *stability*, and *variables of interest*. The complexity of a community is characterized in terms of species richness, their connectance, or interaction strength. The variables of interest are usually species abundances or species biomass. From there, we have several definitions of distinct although associated concepts (see Justus 2008a, 2008b, and 2011 as well).

> STABLE: A system is stable just in the case when all of the variables return to their initial equilibrium values following a perturbation.
> LOCALLY STABLE: A system is locally stable if the return applies only to small perturbations.
> GLOBALLY STABLE: A system is globally stable if the system returns from all possible perturbations.
> RESILIENCE: How fast the variables return to their equilibrium following a perturbation.[44]
> PERSISTENCE: The time a variable lasts before it changes to a new value.
> VARIABILITY: The degree to which a variable varies over time.

At first glance, it appears that ecologists were providing conceptual clarity, and they were recognizing possible hypotheses that were previously unnoticed.

Philosopher Kristin Shrader-Frechette and ecologist E. D. McCoy (1993) argue actually that this is evidence of deep conceptual confusion and ambiguity at the center of community ecology. They write,

> The five meanings as outlined by Pimm (1984), do not even describe all alleged characteristics of communities; some of the meanings refer to the time during which community changes take place, rather to the changes itself, and some of the concepts presuppose different spatial and temporal scales than others...There is no homogeneous class of processes or relationships that exhibit stability or that a community, and there is no single, adequate account of what community or stability is. (McCoy, 1993, 57–58)

Likewise, they contend that different combinations of concepts might lead to different strategies for conservation.

---

[44] The set of values of the variables for which a system returns is its domain of attraction. A system is locally stable if it returns from perturbations that are arbitrarily small and thus has a much smaller domain of attraction than if it was globally stable.

To the degree that ecological concepts (and theories) differ, to the same extent do the strategies and conclusions of applied ecology differ. If our ecological concepts (and theories) are uncertain, then so are our applications. (Shrader-Frechette and McCoy, 1993, 54)

But, a crucial assumption in their argument is that there needs to be a *single* adequate account of what stability is.

To see why we should reject this assumption, we should first notice that these concepts as articulated by Pimm are distinct. Two purportedly different concepts are the same only if they are coextensive. But, it is clear that Pimm's concepts are not coextensive. A community can be locally stable but not globally stable since the former concerns small perturbations and the latter any perturbation. Or, the fact that a community is stable says nothing about how resilient it is; that is, how fast it will return to equilibrium. The resistance of a community is determined by how long the variable of interest maintains a particular value. But, a community's variable of interest can be resistant and unstable.

Of course, Shrader-Frechette and McCoy could agree there are several concepts that are distinct and individually precise and thus not uncertain. They could also claim that there is conceptual confusion because different ecologists have used different terms for the same concept. However, even when this occurs, it is clear their writings often mark the same distinctions. For example, Orians and Pimm both have terms for a variable's return to equilibrium ("trajectory stability" and "stable," respectively), the speed of the return to equilibrium ("elasticity," "resilience"), the time during which a variable maintains the same value ("inertia," "resistance"), the area over which a system returns to equilibrium ("amplitude," "local stability" and "global stability"), and so on. Although different terms are used, the shared meanings of the concepts are clear. Additionally, ecologists like Orians and Pimm do not use the same term to refer to different concepts since they have marked different concepts with different terms. This is true even if the specific terms they use sometimes vary.

Thus, May's models, even if unrealistic, provided resources for ecologists to do their own sort of conceptual analysis and provided resources for new hypotheses to be articulated. May's models, and the conceptual work done analyzing them, has opened up new vistas of models (Allesina and Tang, 2012; Hastings et al., 2016) but also experimental work (Cardinale et al., 2011; Hooper et al., 2012). It is arguable that this experimental work could not have been done without the idealized models.[45]

---

[45] For philosophical analysis of these concepts, see Mikkelson (1997), Odenbaugh (2001), Justus (2008a, 2008b, 2011), and DeLaplante and Picasso (2011). Mikkelson (1997) argues that

One such groundbreaking experimental study, which depended on the conceptual refinement begun in May's work, was done by David Tilman and John Downing (1994). They conducted experiments at Cedar Creek in Minnesota. He and his colleagues introduced seeds from one, two, four, eight, or sixteen grassland-savanna perennial species. The species composition for each treatment was generated at random from an eighteen-species pool. Similar experiments have been done in Europe (Hector et al., 1999; Marquard et al., 2009; Isbell et al., 2011). At Cedar Creek, productivity, as measured by plant biomass, rapidly increased with the addition of species, but tapered off asymptotically. They found that plots with a higher diversity of plant species maintained greater plant productivity during a severe drought than did plots with a lower diversity of plant species.

There are a variety of mechanisms or processes thought to explain why increasing species richness increases productivity. One is *niche complementarity*, which occurs when species differ in how they use limiting resources. If species exploit different resources, then this can increase efficiency and thus higher productivity. Another is the *sampling effect* in which a larger group of species is more likely to contain more productive species that will dominate the community. The main evidence for niche complementarity is trangressive overyielding in which species mixtures yield more biomass than any monocultural plot (Loreau et al., 2001). Additionally, with a positive relationship between species richness and ecosystem functioning, there is also a negative relationship between ecosystem functioning and niche overlap (Wojdak and Mittelbach, 2007). There is some evidence that the sampling effect matters more in the beginnings of an experiment and niche complementarity becomes more important later (Cardinale et al., 2007).

The crucial conceptual point is this. May provided a model that suggested that complexity was inversely proportional to stability. However, with the naturalistic conceptual analysis provided by ecologists like Pimm and Orians, there were other hypothesis to be investigated. Tilman and Downing provided evidence that increasing species richness reduces variation in biomass. This experimental work crucially depended on unrealistic ecological models.

---

notion of Lyapunov stability that May employs makes it unlikely that communities with more species are more likely to be stable by definition. A community is stable only if each species is locally stable. But, the probability that species has any property decreases as the number of species increases. Justus (2008b, 429–430) replies that interspecific interactions that increase with the number of species can actually increase the likelihood of stability. Justus argues that Lyapunov stability is often inappropriate because it holds fixed parameters like r_i, K_i, and a_{ij} that actually change in ecological systems. Thus, we need a "structural stability" concept.

2.4 Conclusion

In this section, I have provided a framework for understanding how models are tested in ecology. We considered an extended example of how models incorporating time delays were evaluated against Nicholson's blowflies. Additionally, we considered models that were untestable, untested, or even testably false, but still were successful for certain purposes.

We now turn out attention to how models intersect with issues in environmental policy and politics more generally.

3 How Ecological Models Inform Environmental Policy and Politics

In this section, I consider two examples of how models have been used in shaping environmental policy and politics. The first model concerns the northern spotted owl, which has been extremely important in environmental debates in the Pacific Northwest of the United States. The second model is the species–area model (SAR) that has been used in a variety of contexts, including island biogeography. Most recently, it has been used to predict future extinction rates. Both models occur alongside and interacted with sociopolitical values, but in different ways.

3.1 Northern Spotted Owls

As we have seen, metapopulation models are important in ecology. They have also been used in one of the most contentious environmental debates.[46] The debate concerned what to do with the northern spotted owl (*Strix occidentalis caurina*) and its habitat (Norse, 1989; Yaffee, 1994; Durbin, 1996). The northern spotted owl is a subspecies that lives only in old-growth coniferous forests in northern California, western Oregon, and western Washington. They are monogamous and each breeding pair lives in 1– sq mi of woodland which is at least 150 years old. Logging and agriculture reduce their habitat, and so managers and policymakers wanted to determine what would prevent them from going extinct. One way of answering that question is to use metapopulation models.

In 1985, environmental lawyer Andrew Stahl presented evolutionary biologist and ecologist Russell Lande with just that question (Yaffee, 1994, 98). Stahl showed Lande the US. Forest Service's "regional guide," which was the plan

---

[46] It is difficult from the outside to appreciate how fractious the debates were during the 1980s. For example, it was not uncommon to see bumperstickers in the Pacific Northwest which said, "Shoot an Owl – Save a Logger." Radical environmental activists such as Earth First! would also chain themselves to old growth trees to protect the trees and habitat.

for conserving the subspecies. He asked Lande if he could determine whether it was up to the task. Lande devised a metapopulation model using a variety of data to provide an answer (Lande, 1988a, but see Lande, 1988b as well). Here is Lande's analysis.

Let ϵ be the probability that a juvenile female inherits her mother's territory, m be the number of territories a juvenile can disperse through before dying, h be the proportion of habitable territory, and p be the proportion of occupied sites. Thus, $(1-\epsilon)$ is the probability that a juvenile doesn't inherit her mother's territory. Additionally, ph is the proportion of habitable but occupied habitats. Lastly, $(1-h)$ is the proportion of inhabitable habitat. Thus, the probability of not inheriting a mother's territory, and to either arrive in an already occupied habitat or unsuitable habitat after m tries is,

$$(1-\epsilon)(ph+1-h)^m \tag{3.1}$$

Therefore,

$$1-(1-\epsilon)(ph+1-h)^m \tag{3.2}$$

is the probability of successfully occupying a suitable habitat after m tries.

Female northern spotted owls reproduce only when they are three years old, and we can represent their growth rate by N_{t+1}/N_t. When $N_{t+1}/N_t = 1$ and there is a stable age distribution, then the population is in a demographic equilibrium. Assuming they are in such an equilibrium, then,

$$[1-(1-\epsilon)(ph+1-h)^m]\,R_0' = 1 \tag{3.3}$$

$R_0' = \sum_{x=0}^{\infty} l_x f_x$ where l_x is the probability of surviving to age x given she has found habitable territory and f_x is the mean lifetime offspring production per female of age x, assuming she finds habitable territory.[47] Finally, solving for $\hat{p}$ we have,

$$\hat{p} = \begin{cases} 1 - \frac{1-k}{h} & \text{for } h > 1-k \\ 0 & \text{for } h \le 1-k \end{cases} \tag{3.4}$$

---

[47] R_0 is the net reproductive rate, which is the average number of female offspring produced per a female over her lifetime. To find R_0, we multiply each value of $l'(x)$ by the corresponding value of $f(x)$ and sum these products across all ages. It represents the reproductive potential of a female during her lifetime adjusted by survivorship. If $R_0 = 1$, then the production of offspring exactly balances mortality. If $R_0 > 1$, then the production of offspring exceeds mortality, and if $R_0 < 1$, then the production of offspring is less than mortality.

In model (3.4), k is the the the equilibrium occupancy of the territory.[48] Model (3.4) implies that there is an "extinction threshold" where $\hat{p} = 0$ if $h = 1 - k$. Thus, persistence occurs when $\hat{p} > 0$, which requires $h > 1 - k$. Lastly, if we assume a population is in demographic equilibrium, then,

$$k = 1 - h(1 - p) \tag{3.5}$$

In 1987, 38 percent of forests in western Washington and Oregon was older than 200 years. So, $h = 0.38$. Fieldwork suggests that 44 percent of sites were occupied. Thus, $p = 0.44$. From (3.5), $k = 0.79$. The forest plans suggested leaving between 7 and 16 percent of forest that is 200 years old or older. But given that $1 - k = 0.21$, Lande argued that 7–16 percent was insufficient and the forest plans should be revised. Lande and others' work led to the US Forest Service withdrawing six old-growth forest timber sales in Oregon and Washington and became a component of the 1994 adoption of the Northwest Forest Plan that protected 8 million acres of old-growth forests. This increase protected 80 percent of the owl's remaining habitat in comparison to 7 percent protected by the Forest Service's original owl plan.

Models like Lande's, and population viability analysis (PVA) more generally, are often criticized (see Beissinger and McCullough (2002); Sarkar (2005) for a discussion of PVA). These models often exhibit parameter or structural sensitivity. If parameter values or assumptions of the models are changed, the models make different predictions. For example, Patrick Foley examined a model of grizzly bears *Ursus arctos* in Yellowstone National Park that incorporated environmental stochasticity along with intrinsic growth rates and carrying capacity (Foley, 1994). The model for the aforementioned estimated values predicted the expected time to extinction was 12, 000 years. However, he later examined a model that included demographic and environmental stochasticity, but set demographic stochasticity to zero (Foley, 1997). The expected time to extinction was approximately fifty years. Thus, PVA models can be very fragile with respect to their parameter values and functional forms. Sensitivity analysis can determine just how much changes in parameter values affect the model's predictions.

Of course, Lande's metapopulation model abstracts and idealizes. For example, he assumes that the northern spotted owl population was approximately in demographic equilibrium. However, he also argues that more realistic models with an Allee effect caused by difficulty finding mates, an edge effect

---

[48] $k = [(1 - 1/R_0')/(1 - \epsilon)]^{1/m}$ and is the "demographic potential" of the metapopulation because it gives the maximum occupancy of suitable habitat $\hat{p} = k$ if $h = 1$.

due to the small extent of suitable habitat, or demographic and environmental stochasticity all reduce k further. Thus, he contended his analysis was conservative.

Lande's analysis was incredibly important for protecting the northern spotted owl. In a interview, he says

> My analysis of the data on spotted owls, which led to nearly a thousand pages of court transcripts of my verbal testimony, showed that the US government plans were likely to drive the Northern Spotted Owl to extinction. It turned out to be the biggest environmental court battle of the decade. I was the key expert witness in the lawsuits against three branches of the US government responsible for management of wildlife that were brought up by conservationists. (Fisch, 2012, 24)

Lande's models and expert testimony were used in the service of litigating the US government to protect the northern spotted owl. Although this is unusual in certain respects, it is not morally problematic since often scientists can and should provide advice to policymakers.[49]

3.2 Species Area Models

Many ecologists and conservation biologists argue we are on the cusp of the sixth mass extinction.[50] On the basis of several lines of evidence, they claim that current rates of species extinction are very high and are far greater than the "normal" background rates as found in the fossil record. There are three general approaches to projecting extinction rates. First, there is the approach that utilizes species area models, which we will consider. Second, there is an approach that examines how well-studied taxa move through the categories *vulnerable*, *endangered*, *probably extinct*, and *certified extinction* on the International Union for the Conservation of Nature and Natural Resources' (IUCN's) "red lists." Third, there is an approach that uses IUCN data to estimate the probability of extinction as a function of time. Nevertheless, if we are to

---

[49] For example, in the Ecological Society of America's Code of Ethics, www.esa.org/about/code-of-ethics/, it states,

> Ecologists will, to the extent practicable, engage meaningfully with the communities in which they practice to promote teaching, learning and an understanding of their study; broaden the participation of underrepresented groups; enhance local infrastructure for research and education; and disseminate results broadly to benefit the local community.

Of course, one should use models only when they are well-evidenced for the purposes to which they are put. Lande's models were taken to be well-supported at the time for making these sorts of predictions. Nevertheless, the advice they provide is only as good as the models themselves, and courts lack the scientific expertise to evaluate them.

[50] See Myers (1979), Ehrlich and Ehrlich (1981), and Wilson (1988, 1992).

determine that we are in a biodiversity crisis, then we must determine three rates: the background extinction rate, the current extinction rate, and the projected extinction rate. If the projected rate is much greater than that of the background rate, then we have reason to believe we are in an extinction crisis.[51]

The first argument given by biologists essentially depends on species area models used in community ecology. It is this form of argument that I discuss in this section. The species-area argument goes roughly like this: There are between 5 and 30 million extant species on the Earth. Tropical rainforests are being destroyed at approximately 2 percent per year. On the basis of the SAR, the number of species being lost per year greatly exceeds that of the background rate of the fossil record. In the fossil record, approximately one species would be lost per year, and our model predicts that between 10, 000 and 27,000 species will be lost per year. Hence, the expected number of human-caused species extinctions is much greater than that before our appearance. It is then suggested that we are in an extinction crisis.

The species area relation has a long legacy in ecology. Ecologists have long known that, as area increases, so does the number of species in that area. The biogeographer Phillip Darlington proposed that, with every tenfold increase in area, the number of species doubles.[52] Michael Rosenzweig writes, "You will find more species if you sample a larger area. That rule has more evidence to support it than any other about species diversity" (Rosenzweig, 1995, 8). Ecologist Nicholas Gotelli claims that the species–area relationship is one of the few laws in ecology (Gotelli, 1995, 172).

One SAR representing this relation that fits many data sets is the power function,

$$S = cA^z \tag{3.6}$$

where c is a fitted constant, and z is a parameter that generally has values specific to the type of area under consideration. We can transform equation (3.6) logarithmically into the following,

$$\log(S) = \log(cA^z)$$
$$= \log(c) + \log(A^z)$$
$$= \log(c) + z\log(A)$$

---

[51] It is worth noting that "crisis" is a normative term (Sarkar, 2005, 134–135). Hence, the claim that we are in an extinction crisis is a normative claim. Sahotra Sarkar's own suggestion is there is biodiversity crisis if the current extinction rate is twice the background extinction rate.

[52] However, it should be noted that Darlington was not the first to suggest this rule. Rather, H. G. Watson in 1835 remarked that, in England, as the area of a county increases by 10, the number of plant species increases by 2 (Connor and McCoy, 2001, 397).

The transformation gives us a family of straight lines where $\log(c)$ is the y-intercept and z is the slope of the line. If $z = 0.3$, then we can derive Darlington's rule.

Ecologists Robert May, John Lawton, and Nigel Stork (1995)'s argument for an extinction crisis is based on the following approximation. We have a linear approximation to the species–area curve at the point $(S_{original}, A_{original})$ with the following equation.

$$\Delta S = z\Delta A \tag{3.7}$$

TECHNICAL DISCUSSION

Equation (3.7) approximates equation (3.6) at the point $(A_{original}, S_{original})$. In general, if we want to approximate a curve at a point, we find the tangent line at that point on the curve. The tangent line has the same direction and slope as the curve itself. Hence, near the point of interest, it has a similar behavior. First, we take the derivative of $S = cA^z$ in order to find the slope of the tangent line.

$$\frac{dS}{dA} = \frac{d(cA^z)}{dA}$$
$$= \frac{A^z dc}{dA} + \frac{cdA^z}{dA}$$
$$= \frac{czA^{z-1}}{dA}$$

Next we multiply both sides by dA, giving us,

$$dS = czA^{z-1}da$$

We now divide both sides by S, resulting in,

$$\frac{dS}{S} = \frac{czA^{z-1}dA}{S}$$

Substituting cA^z for S on the right-hand side gives us,

$$\frac{dS}{S} = \frac{czA^{z-1}dA}{cA^z} = z\left(\frac{dA}{A}\right)$$

Now dS and dA are infinitesimally small changes in S and A. if we consider larger changes in $A_{new} \ll A_{original}$, we have the approximation, which is $\Delta S = z\Delta A$.

---

where ΔS is the proportional reduction in species richness, and ΔA is the proportional reduction in area. Their argument applies this model (3.7) to global losses of species with a conservative value of $z = 0.25$. Rates of tropical deforestation have been claimed to be in the range of 0.8 − 2% per year. If $\Delta A = 0.8\%$, then our minimal prediction is $\Delta S = 0.2\%$. If $\Delta A = 0.2\%$, then our maximal prediction is $\Delta S = 0.5\%$. So, between 0.2 and 0.5 percent of species will go extinct per year. May et al. suppose conservatively that there are 5 million species globally. Thus, our model and estimates project a minimal loss of 10,000 species per year and projects a maximal loss of 25,000 species per year. This is equivalent to an average species' lifespan between 200 and 500 years. Finally, this is also equivalent to losing approximately one to three species per hour.

Why would one be skeptical of this argument? It is crucial to note that there are a variety of uncertainties involved in our species area argument. First, our model is idealized since we are assuming that loss of species is only a result of loss of area, and thus we are ignoring many important causal factors related to extinction like invasives, disease, overhunting, habitat fragmentation, edge effects, and habitat diversity. Second, we do not know exactly how much habitat we are losing per year. At best, we are losing between somewhere between 10 million and 15 million ha of closed tropical rainforest per year (and tropical rainforests contain approximately half of the species that exist) (Stork, 2010). Third, we do not know how many species currently exist within an order of magnitude (Erwin, 1982; May, 1990, 2011; Stork, 1993; Mora et al., 2011). There may be as many as 5 − 30 million extant species, given our best evidence and our very poor knowledge of taxa other than mammals, birds, and some insects. There are other objections to this argument that are given.

First, there are data sets that (3.6) do not fit well (Sarkar, 2005). Although true, there are many data sets at a variety of spatial scales such as provinces, archipelagos, and across provinces for which it does fit reasonably well. For example, Triantis et al. (2012) examined twenty models with 601 data sets using AIC and determined that the power model and family were the best-fitting models. As another example, Pimm and Askins (1995) analyzed extinctions in the northeast United States due to deforestation. Suppose the original habitat area is A_0, which is reduced to A_n and the original number of species S_0 is reduced to S_n. Using (3.6), we have $S_n/S_0 = cA_n^z/cA_0^z = (A_n/A_0)^z$. They estimate that there has been a 50 percent reduction in forest area from 1620 to 1873, where $z = 0.25$. Thus, $(0.5)^{0.25} \approx 0.84$, which is the proportion of species that remain. Therefore, given that there were 160 bird species in the original forest, 16 percent of twenty-six species should have gone extinct. Historical evidence suggests only five species have gone extinct. Pimm and Askins noted that the model applies specifically to endemic species, of which there were twenty-eight. Of those, 16 percent of those should have gone extinct, which is 4.5 and very close to the observed extinctions.

Second, the causes of the species–area relationship are much debated (Connor and McCoy, 1979, 2001). For example, does the species–area relationship result from larger areas per se because they can support more species? Is it that larger areas contain more habitats and greater number of habitats supports more species? Is it that larger areas are more likely to receive more colonists than smaller areas? Finally, do larger areas contain a greater number of resources and thus larger areas support more species? The causes generating species–area relationships – be it area per se, habitat diversity, passive sampling, or resource concentration – are of great importance. However, first we should distinguish skepticism concerning what *process* generates the species–area relationship versus the *pattern* itself. Second, the pattern is a positive monotonic relation between area and species diversity. There are a variety of functional forms consistent with the species–area relationship, even if no single one fits every data set well.[53]

Some have argued that the exact functional form of the species–area relationship doesn't matter. W. V. Reid writes, "it can be argued that the exact rate of extinction is not terribly important given that current extinction rates greatly exceed background rates" (Reid, 1992, 55). Daniel Simberloff writes, "Nevertheless, it is a worthwhile exercise to use the species–area relationship to attempt a first guess at how many extinctions deforestation will generate in tropical forests" (Simberloff, 1992, 78).

Third, the derivation of (3.7) does not by itself justify how it is often used. It is derived by finding a linear approximation to (3.6) at a point. However, proponents of the species area argument do not restrict their analysis to such nearby points. They extrapolate to much, much larger reductions in area. Thus, the linear approximation will not necessarily hold, and thus habitat reductions might not behave in accordance with (3.6). It *might*, but this has to be shown by data. Put differently, it is quite possible that, with small reductions of habitat, the reduction in species will behave in accordance with $S = cA^z$, but will not for much larger reductions.

Fourth, some have argued that, since the equilibrium model of island biogeography (MacArthur and Wilson, 1967) has been seriously criticized, we should be similarly skeptical of the species–area relationship. However, the species–area relationship and the curves that model it are distinct from the equilibrium theory of island biogeography. The latter attempted to explain such

---

[53] Even if we do not know what causes give rise to the species–area relationship, the relationship itself could be causal. Following James Woodward (2003), a generalization describes a causal relationship if it is invariant under a set of interventions. Other things being equal, if we intervene on A and there is a resulting change in S, we have a causal relationship.

relationships by examining how rates of immigration to islands from the mainland and rates of extinction on islands lead to steady state diversities. However, we should not confuse a purported explanation of a pattern with the pattern itself.

Michael Rosenzweig has argued that the aforementioned species–area argument is optimistic (Rosenzweig, 2003, 199). Note the SAR represents any type of area; it is the z-value specifies the specific type of area. The z-values as measured empirically depend on the scale one is considering. Sample area–species area relationships have a z-value between 0.1 and 0.2, archipelagic species–area relationships have z-values between 0.25 and 0.55, and interprovincial species area relationships have z-values between 0.6 and 1.0 (Rosenzweig, 2003, 195–196). So, let us just consider the proportion of species remaining as determined by the proportion of area remaining. That is, given S' and S are the new and old number of species respectively and A' and A are the new and old amount of area respectively, then,

$$\left(\frac{S'}{S}\right) = \left(\frac{cA'^z}{cA^z}\right) = \left(\frac{A'}{A}\right)^z \tag{3.8}$$

If we suppose the Earth is like an archipelago as May et al. do, then we could argue, for example, that a reduction to 2 percent area with a $z = 0.25$ would leave 38 percent of the species remaining (i.e., $0.02^{0.25} \approx 0.38$). However, in looking at global losses of species, we should not be looking at archipelagos but provinces. As Rosenzweig notes,

> A biogeographical province is a region whose species have evolved within it, rather than immigrating from somewhere else. Although the concept is merely an ideal – every place has at least a few species that arrived as immigrants – it is close to true in many places, such as different continents or well-separated periods in the history of life. (Rosenzweig, 2003, 195–196)

Thus, for provinces, their steady state derives from their rate of speciation and rate of extinction. He goes on,

> The world of nature reserves is not an island but a shrunken province. Its source pool is the past. Species that become extinct in it cannot immigrate from the past to recolonize the world of the future. So, like any evolutionarily independent providence, our miniaturized natural world must seek its steady state along the interprovincial SPAR, not the island SPAR. (Rosenzweig, 2003, 200)

The appropriate z-value is then between 0.6 and 1.0. Given this model, we have for a reduction to 2 percent area and $z = 0.8$, and so 4 percent of species are remaining (i.e., $0.2^{0.8} \approx 0.04$). Given more accurate considerations of z-values and our SAR, it looks like matters could be far, far worse.

3.3 Ecological Models and Inductive Risk

E. O. Wilson writes regarding biodiversity loss and the SAR,

> There is no way to measure the absolute amount of biological diversity vanishing year by year in rain forests around the world, as opposed to percentage losses, even in groups as well known as the birds. Nevertheless, to give an idea of the dimension of the hemorrhaging, let me provide the most conservative estimate that can be reasonably based on our current knowledge of the extinction process. I will consider only species being lost by reduction in forest area, taking the lowest z value permissible (0.15). I will not include overharvesting or invasion by alien organisms. I will assume a number of species living in the rain forests, 10 million (on the low side), and I will further suppose that many of the species enjoy wide geographical ranges. Even with these cautious parameters, selected in a biased manner to draw a maximally optimistic conclusion, the number of species doomed each year is 27,000. Each day it is 74, and each hour 3. If past species have lived on the order of a million years in the absence of human interference, a common figure for some groups documented in the fossil record, it follows that the normal "background" extinction rate is about one species per one million species a year. Human activity has increased extinction between 1,000 and 10,000 times over this level in the rain forest by reduction in area alone. Clearly we are in the midst of one of the great extinction spasms of geological history. (Wilson, 1992, 280)

Of course, Wilson knows the tremendous uncertainties around estimating rates of tropical deforestation, estimating the background rate of extinction, and the debates over model (3.6). So, how can he be so confident with are in "great extinction spasm"? He continues,

> In order to set a lower limit above which the species extinction rate can be reasonably placed, I will employ what we know about the relation between the area of habitats and the numbers of species living within them. Models of this kind are used routinely in science when direct measurements cannot be made. They yield first approximations that can be improved stepwise as better models are devised and more data added. (Wilson, 1992, 275)

Essentially, he assumes that (3.6) provides a "lower limit" to human-caused species extinction. Recent critics Fangliang He and Stephen Hubbell (2011) controversially think that (3.6) actually overestimates species extinction.

He and Hubbell argue that the predictions of species extinctions from (3.6) lack empirical support. A common response to why we don't see as many extinctions as predicted is because of "extinction debt." In effect, after habitat is destroyed, there is a time lag between when habitat is destroyed and when the species finally goes extinct (Tilman et al., 1994). However, He and Hubbell argue that (3.6) overestimates species extinctions by assuming that the sampling problem for extinction is just the reverse of sampling for the species–area relationships. They claim that the area in which one finds the first individual of a species is much smaller than the area that must be eliminated to extinguish the last individual of a species. Thus, in general, it takes a larger loss of area to drive a species extinct than to add the species. He and Hubbell devise a SAR by adding a new species every time the sampling frame a encounters the first individual of that species. They devise the endemic area model (EAR) by adding a species when all of the individuals of the species are in a. If we assume species are randomly and independently distributed, then the SAR curve is,

$$S_a^1 = S - \sum_{i=1}^{S} \left(1 - \frac{a}{N}\right)^{N_i} \tag{3.9}$$

The EAR curve is,

$$S_a^N = \sum_{i=1}^{S} \left(\frac{a}{A}\right)^{N_i} \tag{3.10}$$

where S is the number of species in A and N is the abundance of a given species in A. Now, suppose the total area is A and a subarea a is lost. If the species are randomly and independently distributed in space, then the expected number of species extinct from a loss of a according to (3.9) is $S_{\text{loss}} = S - S_{A-a}$. However, it turns out that $S_{\text{loss}} = S_a^N$. Thus, when species are randomly and independently distributed, then SAR and EAR agree. Under random and independent placement, the area to find the first individual a^1 and the area to find the last individual a^N are equal to A; that is, $a^1 + a^N = A$. However, with a clumped distribution, $a^1 + a^N < A$ and $a^N \geq a^1$. Therefore, when species are clumped, then $S_{\text{loss}} \neq S_a^N$. This is why He and Hubbell claim the backward SAR overestimates extinction rates.

He and Hubbell's contentious work has been subject to many criticisms and responses (Brooks, 2011; Evans et al., 2011; He and Hubbell, 2012, 2013; Pereira et al., 2012; Thomas and Williamson, 2012; Axelsen et al., 2013). The first criticism is that (3.6) fits the data far better than He and Hubbell suggest, as we discussed in the previous section. The second is that their models contain "analytic errors." Axelsen et al. (2013) point out that EAR can be derived from SAR according to the following,

$$\text{EAR}(a) = s - \text{SAR}(A - a) \tag{3.11}$$

where $SAR(a) = \sum_{j=1}^{s} p_j(a)$ and $p_j(a)$ is the probability that species j is contained in subarea a and $EAR(a) = \sum_{j=1}^{s} q_j(a)$, where $q_j(a)$ is the probability species, and j is endemic to a. He and Hubbell claim that "[o]nly in a very special and biologically unrealistic case" when species are arranged randomly and independently in space can one derive EAR from SAR (He and Hubbell, 2011, 368). However, Axelsen et al. (2013) argue that this must be incorrect. The event that j is endemic to a is equivalent to the event that is not contained in $A - a$. The probability of the former is $q_j(a)$ and the latter is $1 - p_j(A - a)$. Hence, $q_j(a) = 1 - p_j(A - a)$ (for a response, see He and Hubbell (2013)).

Interestingly, some of the most intense criticisms were to be found in a piece "Scientists Clash on Claims Over Extinction 'Overestimates'" in the New York Times.[54] Hubbell himself thought the claim that SAR overestimates extinctions was in a way good. He said, "This is welcome news in that we have bought a little time for saving species. But we have to redo a whole lot of research that was done incorrectly." However, other ecologists reacted very differently. Michael Rosenzweig said, "I was desperately upset by it. I can't begin to tell you." Stuart Pimm said the paper was an "outrage." He continued, "They say they've got a recipe for baking bread. What they've done is bake cookies." He suggested that, if it didn't have the inflammatory title, He and Hubbell's paper would not have been published in *Nature*. Additionally, he chastised them for not reading his own paper (although they cited it) because his predictions were correct. Most interestingly, Daniel Simberloff worried about the political implications of the paper. He said, "This sounds like a major argument against preserving large areas, and in practical terms it shouldn't be." Regarding the projected extinction rate being larger than the background extinction rate, "Does it really matter if it's 100 times or 1,000 times?"[55] Ann Kinzig suggested, "There's enough going on in the world that we shouldn't be over-alarmist about this. Let's try and get it right." Philosophers of science shouldn't be surprised at the rancor in this or other scientific disputes. But, some would argue the heightened emotions and sharp tongues are irrelevant to the issues of the previous section. The political implications of rejecting the SARs are irrelevant to whether they are approximately true or fit the data. I now want to consider an argument for why those implications might be relevant to model testing.

The argument I will sketch concerns the problem of inductive risk (Rudner 1953; Hempel 1965b; Douglas 2000, and, in the context of ecology, see

---

[54] https://archive.nytimes.com/www.nytimes.com/gwire/2011/05/18/18greenwire-scientists-clash-on-claims-over-extinction-ove-96307.html?pagewanted=3.

[55] Pereira et al. (2012) point out that the differences between SAR and EAR exist but are negligible when considering the problem of species extinction.

Table 3.1 Type I and Type II errors.

	H_0 is true	H_0 is false
Accept H_0	Correct	Type II error
Reject H_0	Type I error	Correct

Shrader-Frechette and McCoy, 1993, ch. 6). In standard Neyman–Pearson hypothesis testing we formulate a null hypothesis H_0 and an alternate H_1 (Gotelli and Ellison, 2012). Schematically, H_0 states that, "Cause C does not have an effect E." H_1 states that "Cause C has an effect E." Thus, there are two relevant probabilities of error. To use the example of this section,

> H_0: Habitat destruction will not severely increase species extinctions
> H_1: Habitat destruction will severely increase species extinctions

Here are our two errors. We reject H_0, and habitat will not severely increase species extinctions. We accept H_0, and habitat destruction will severely increase species extinctions. We cannot minimize both Type I and Type II errors.[56] Which would be worse – allowing habitat destruction although it increases species extinctions or preventing habitat destruction although it will not alleviate species extinctions? We have a moral and a sociopolitical question. Given habitat destruction affects both species extinctions and human well-being, this is not merely a technical scientific question.

The argument from inductive risk in its general form goes something like this (Steel, 2010). One important goal of scientific inference is to decide whether to accept or reject hypotheses. Our decisions as whether to accept or reject a hypothesis should depend in part on various value judgments about the consequences of accepting the hypothesis if false and rejecting it if true. Thus, values should influence scientific inference. Hence, in application to the current discussion, we should decide whether to accept or reject our SAR based in part on potential harms habitat destruction and the attending species extinction would cause.

Let me offer three responses to the argument. First, suppose the argument is sound. It might follow that we should accept the hypothesis that habitat destruction will severely increase species extinctions. But, this does not imply we should accept (3.6) as approximately true or predictively accurate. Second, accepting a hypothesis might involve *believing* the hypothesis is true or *acting*

---

[56] One of my philosophical heroes William James talked of the tension between "Believe truth!" and "Shun error!" and wrote, "For my own part, I have also a horror of being duped; but I can believe that worse things than being duped may happen to a man in this world" (James 1979).

as if the hypothesis is true (van Fraassen, 1980).[57] Thus, the consequences of habitat destruction might give us reason to act as if (3.6) is approximately true, but not believing it true or predictively accurate.[58] Third, the Neyman–Pearson framework is a very common statistical methodology. However, some philosophers and statisticians reject it for Bayesianism (Howson and Urbach, 2006). According to Bayesianism, we neither accept nor reject hypotheses. Rather, we compare the posterior probability $\Pr(H|E)$ of hypothesis with regard to some evidence and its prior probability $\Pr(H)$, updating according to $\Pr_{new}(H) = \Pr_{old}(H|E)$. The notion of Type I and Type II errors depends on the notion of hypothesis acceptance and rejection as opposed to changing credences in light of evidence. Thus, Bayesianism doesn't involve acceptance of hypotheses at all.[59]

Scientific decision-making involves values. Some philosophers argue that the only values present in the science are epistemic values (Laudan, 1986). Thomas Kuhn argued that scientific theories are evaluated in terms of accuracy, consistency, scope, simplicity, and frutifulness (Kuhn, 1977). Additionally, he said these criteria are values because they require interpretation given their vagueness, and they require weightings given there is no lexical priority. Ecology obviously includes these epistemic values. However, it is a complicated science in part because moral and sociopolitical values often are so close at hand.

3.4 Conclusion

In this section, we have considered two models that bear on environmental policy and politics. First, we considered Lande's metapopulation model, which he devised as practical guidance for determining how much habitat was needed to conserve the northern spotted owl. Second, we examined SARs and their predictions for species extinctions due to habitat destruction. These models are much more controversial than Lande's. However, these models are associated with irreversable harmful effects on nonhuman life and human well-being. How

---

[57] However, as Horwich (1991) has argued, belief and acceptance may come to the same thing functionally or pragmatically. Suppose we accept a model to make predictions, to come to decisions, and provide explanations. How is this not believing that the model is roughly right?

[58] Following moral consequentialism, we might embrace epistemic consequentialism, which says we should believe that which has the best expected consequences.

[59] Critics of Bayesianism do offer a few responses. First, there are other probabilities involved in scientific inference than the ones mentioned earlier and they involve inductive risk too (Douglas, 2009). Here the most obvious examples are estimates of uncertainty. Second, often we lack detailed information regarding credences, and thus scientists must decide how to represent these probabilities with a distribution (e.g., normal, binomial, Poisson, etc.) that introduces inductive risk (Steel, 2013). Third, one might work out a notion of acceptance in the Bayesian framework (Levi, 1967; Maher, 1993)

to think about the roles those values play in model acceptance and rejection is something we only broached. It deserves more attention than we could give it here.

4 Conclusion

In this Element, we considered three issues: What are ecological models? How are ecological models tested? How do models inform environmental policy and politics? Although we could only scratch the proverbial surface, I hope this has excited you to dig deeper into extremely exciting philosophy and ecology. These models bear on very practical issues existentially important to us of all.

Of necessity, many issues have been left open-ended, which I would explore further if for more space. But, let me mention just a few. First, I sketched – and really just sketched – a deflationary approach to models and modeling. The similarity approach has been worked out in detail, and this competitor needs the same treatment. How can we bring resources regarding linguistics, aesthetics, and other disciplines to help us understand these representations? Second, it is clear, I hope, that models serve as important heuristics even when they're unrealistic and predictively inaccurate. However, there is a legitimate worry that this defense proves too much. Exactly when are models useful as heuristics, and when are they not? Here we need more work in the cognitive science of modeling. Many philosophers and ecologists talk of model building as a "strategy," but effective strategies require a keen analysis of when they will succeed and when they won't. We haven't provided the kind of engineering analysis that is required. Third, it is certainly true that there can be too many or too few modelers. What is the optimal distribution between modelers, experimentalists, and field ecologists? How can we ensure that models are adequately tested? What sorts of incentive structures can be placed in undergraduate and graduate ecology programs that create this balance? Fourth, ecology with its models is coopted into other very practical disciplines like conservation biology and environmental studies. For example, simple ecological models, like the logistic people tell us, imply drastic population reduction is required to "save the Earth." But, such practical disciplines are rife with sociopolitical values, and it is not clear how much work the ecological models really are doing. Are they politics just in a different name? How do values interact, and how should they, with ecological models?

The philosopher Sir Karl Popper once wrote,

> In my opinion, the greatest scandal of philosophy is that, while all around us the world of nature perishes – and not the world nature alone – philosophers

continue to talk, sometimes cleverly and sometimes not, about the question of whether this world exists. (Popper, 1979, 32)

Although I am sympathetic with Popper's point, if we are to prevent nature's perishing, we must consider its reality. Or, at the very least, whether sciences like ecology represent this reality accurately. This Element has been an examination of how we can do this. For, it is ecology, along with much else of course, which can help us curtail our destructive tendencies and care for our more-than-human world.

Bibliography

Allesina, S. and S. Tang (2012). Stability criteria for complex ecosystems. *Nature 483*(7388), 205.

Andrewartha, H. G. (1954). *The Distribution and Abundance of Animals*. University of Chicago Press.

Axelsen, J. B., U. Roll, L. Stone, and A. R. Solow (2013). Species–area relationships always overestimate extinction rates from habitat loss: comment. *Ecology 94*(3), 761–763.

Bailer-Jones, D. M. (2009). *Scientific Models in Philosophy of Science*. University of Pittsburgh Press.

Beatty, J. (1980). Optimal-design models and the strategy of model building in evolutionary biology. *Philosophy of science 47*(4), 532–561.

Beatty, J. (1982). What's wrong with the received view of evolutionary theory? In *PSA: Proceedings of the Biennial Meeting of the Philosophy of Science Association*, Volume 1982, pp. 397–426. Philosophy of Science Association.

Beatty, J. (1997). Why do biologists argue like they do? *Philosophy of Science 64*, S432–S443.

Beissinger, S. R. and D. R. McCullough (2002). *Population Viability Analysis*. University of Chicago Press.

Bender, E. A. (1978). *An Introduction to Mathematical Modeling*. Dover Publications.

Bernstein, R. (2003). *Population Ecology: An Introduction to Computer Simulations*. John Wiley & Sons.

Berryman, A. A. (2003). On principles, laws and theory in population ecology. *Oikos 103*(3), 695–701.

Bodenheimer, F. S. (1928). Welche faktoren regulieren die individuenzahl einer insektenart in der natur. *Biologisches Zentralblatt 48*, 714–739.

Bogen, J. and J. Woodward (1988). Saving the phenomena. *The Philosophical Review 97*(3), 303–352.

Brandon, R. N. (1997). Does biology have laws? The experimental evidence. *Philosophy of Science 64*, S444–S457.

Brooks, T. (2011). Extinctions: Consider all species. *Nature 474*(7351), 284.

Brown, D., and P. Rothery (1993). *Models in Biology: Mathematics, Statistics and Computing*. John Wiley & Sons.

Burnham, K. P. and D. R. Anderson (2003). *Model Selection and Multimodel Inference: A Practical Information-Theoretic Approach*. Springer Science & Business Media.

Callender, C. and J. Cohen (2006). There is no special problem about scientific representation. *Theoria 21*(1), 67–85.

Cardinale, B. J., K. L. Matulich, D. U. Hooper, et al. (2011). The functional role of producer diversity in ecosystems. *American Journal of Botany 98*(3), 572–592.

Cardinale, B. J., J. P. Wright, M. W. Cadotte, et al. (2007). Impacts of plant diversity on biomass production increase through time because of species complementarity. *Proceedings of the National Academy of Sciences 104*(46), 18123–18128.

Carlson, A. (2000). The effect of habitat loss on a deciduous forest specialist species: The white-backed woodpecker (*Dendrocopos leucotos*). *Forest Ecology and Management 131*(1–3), 215–221.

Cartwright, N. (1983). *How the Laws of Physics Lie*. Oxford University Press.

Cartwright, N. (1994). *Nature's Capacities and Their Measurement*. Oxford University Press.

Case, T. J. (2000). *An Illustrated Guide to Theoretical Ecology*. Oxford University Press.

Castle, D. G. (2001). A semantic view of ecological theories. *Dialectica 55*(1), 51–66.

Chang, C.-W., M. Ushio, and C.-h. Hsieh (2017). Empirical dynamic modeling for beginners. *Ecological Research 32*(6), 785–796.

Colyvan, M. and L. R. Ginzburg (2003). Laws of nature and laws of ecology. *Oikos 101*(3), 649–653.

Connor, E. F. and E. D. McCoy (1979). The statistics and biology of the species–area relationship. *The American Naturalist 113*(6), 791–833.

Connor, E. F. and E. D. McCoy (2001). Species–area relationships. *Encyclopedia of Biodiversity 5*, 397–411.

Cooper, G. (1998). Generalizations in ecology: A philosophical taxonomy. *Biology and Philosophy 13*(4), 555–586.

Cooper, G. J. (2003). *The Science of the Struggle for Existence: On the Foundations of Ecology*. Cambridge University Press.

Costantino, R., R. Desharnais, J. Cushing, and B. Dennis (1997). Chaotic dynamics in an insect population. *Science 275*(5298), 389–391.

Cummins, R. (1989). *Meaning and Mental Representation*. MIT Press.

Currie, A. (2018). Bottled understanding: The role of lab-work in ecology. *The British Journal for the Philosophy of Science*, axy047.

Cushing, J. M., R. F. Costantino, B. Dennis, R. Desharnais, and S. M. Henson (2002). *Chaos in Ecology: Experimental Nonlinear Dynamics*, Volume 1. Elsevier.

Davidson, J. and H. Andrewartha (1948). Annual trends in a natural population of *Thrips imaginis* (Thysanoptera). *The Journal of Animal Ecology 17*(2), 193–199.

De Angelis, D. L. (1975). Stability and connectance in food web models. *Ecology 56*(1), 238–243.

DeAngelis, D. L. and S. Yurek (2015). Equation-free modeling unravels the behavior of complex ecological systems. *Proceedings of the National Academy of Sciences 112*(13), 3856–3857.

DeLaplante, K. and V. Picasso (2011). The biodiversity-ecosystem function debate in ecology. In K. deLaplante, B. Brown, and K. A. Peacock (Eds.), *Philosophy of Ecology*, pp. 169–200. North-Holland.

Douglas, H. (2000). Inductive risk and values in science. *Philosophy of Science 67*(4), 559–579.

Douglas, H. (2009). *Science, Policy, and the Value-Free Ideal*. University of Pittsburgh Press.

Downes, S. M. (1992). The importance of models in theorizing: A deflationary semantic view. In *PSA: Proceedings of the Biennial Meeting of the Philosophy of Science Association*, Volume 1992, pp. 142–153. Philosophy of Science Association.

Dretske, F. (1997). *Naturalizing the Mind*. MIT Press.

Durbin, K. (1996). *Tree Huggers: Victory, Defeat & Renewal in the Northwest Ancient Forest Campaign*. Mountaineers.

Earman, J. (1992). *Bayes or Bust? A Critical Examination of Bayesian Confirmation Theory*. MIT Press.

Egerton, F. N. (1973). Changing concepts of the balance of nature. *The Quarterly Review of Biology 48*(2), 322–350.

Ehrlich, P. and A. Ehrlich (1981). *Extinction: The Causes and Consequences of the Disappearance of Species*. Random House.

Elgin, C. Z. (2004). True enough. *Philosophical issues 14*(1), 113–131.

Elgin, C. Z. (2017). *True Enough*. MIT Press.

Erwin, T. L. (1982). Tropical forests: Their richness in Coleoptera and other arthropod species. *The Coleopterists Bulletin 36*(1), 74–75.

Etienne, R. S. (2002). A scrutiny of the levins metapopulation model. *Comments on Theoretical Biology 7*, 257–281.

Evans, M., H. Possingham, and K. Wilson (2011). Extinctions: Conserve not collate. *Nature 474*(7351), 284.

Fisch, F. (2012). Who's going to speak up for nature? *Lab Times* (1), 2–25.

Foley, P. (1994). Predicting extinction times from environmental stochasticity and carrying capacity. *Conservation Biology 8*(1), 124–137.

Foley, P. (1997). Extinction models for local populations. In I. A. Hanski and M. E. Gilpin (Eds.), *Metapopulation Biology: Ecology, Genetics, and Evolution*, pp. 215–246. Academic Press.

Forster, M. and E. Sober (1994). How to tell when simpler, more unified, or less ad hoc theories will provide more accurate predictions. *The British Journal for the Philosophy of Science 45*(1), 1–35.

Forster, M. R. (2000). Key concepts in model selection: Performance and generalizability. *Journal of Mathematical Psychology 44*(1), 205–231.

Forster, M. R. (2002). Predictive accuracy as an achievable goal of science. *Philosophy of Science 69*(S3), S124–S134.

Gardner, M. R. and W. R. Ashby (1970). Connectance of large dynamic (cybernetic) systems: Critical values for stability. *Nature 228*(5273), 784.

Giere, R. N. (1988). *Explaining Science: A Cognitive Approach*. University of Chicago Press.

Giere, R. N. (1999). *Science without Laws*. University of Chicago Press.

Giere, R. N. (2010). *Scientific Perspectivism*. University of Chicago Press.

Goodman, N. (1968). *Languages of Art: An Approach to a Theory of Symbols*. Hackett Publishing.

Gotelli, N. J. (1991). Metapopulation models: The rescue effect, the propagule rain, and the core-satellite hypothesis. *The American Naturalist 138*(3), 768–776.

Gotelli, N. J. (1995). *A Primer of Ecology*. Sinauer Associates, Inc.

Gotelli, N. J. and A. M. Ellison (2012). *A Primer of Ecological Statistics*. Sinauer Associates, Inc.

Gotelli, N. J. and W. G. Kelley (1993). A general model of metapopulation dynamics. *Oikos 68*(1), 36–44.

Grice, H. P. (1991). *Studies in the Way of Words*. Harvard University Press.

Griesemer, J. R. (1990a). Material models in biology. In *PSA: Proceedings of the Biennial meeting of the Philosophy of Science Association*, Volume 1990, pp. 79–93. Philosophy of Science Association.

Griesemer, J. R. (1990b). Modeling in the museum: On the role of remnant models in the work of Joseph Grinnell. *Biology and Philosophy 5*(1), 3–36.

Gurney, W., S. Blythe, and R. Nisbet (1980). Nicholson's blowflies revisited. *Nature 287*, 17–21.

Hacking, I. (2016). *Logic of Statistical Inference*. Cambridge University Press.

Harrison, S., D. D. Murphy, and P. R. Ehrlich (1988). Distribution of the bay checkerspot butterfly, *Euphydryas editha bayensis*: Evidence for a metapopulation model. *The American Naturalist 132*(3), 360–382.

Hastings, A. (1997). *Population Biology: Concepts and Models*. Springer Science & Business Media.

Hastings, A., C. L. Hom, S. Ellner, P. Turchin, and H. C. J. Godfray (1993). Chaos in ecology: Is mother nature a strange attractor? *Annual Review of Ecology and Systematics 24*(1), 1–33.

Hastings, A., K. S. McCann, and P. C. de Ruiter (2016). Introduction to the special issue: Theory of food webs. *Theoretical Ecology 9*(1), 1–2.

He, F. and S. Hubbell (2013). Estimating extinction from species–area relationships: Why the numbers do not add up. *Ecology 94*(9), 1905–1912.

He, F. and S. P. Hubbell (2011). Species–area relationships always overestimate extinction rates from habitat loss. *Nature 473*(7347), 368.

He, F. and S. P. Hubbell (2012). He and Hubbell reply. *Nature 482*(7386), E5.

Hector, A., B. Schmid, C. Beierkuhnlein, et al. (1999). Plant diversity and productivity experiments in European grasslands. *Science 286*(5442), 1123–1127.

Hempel, C. (1965a). *Aspects of Scientific Explanation*. Free Press.

Hempel, C. (1965b). Science and human values. In *Aspects of Scientific Explanation and Other Essays in the Philosophy of Science*, pp. 81–96. The Free Press.

Hesse, M. (1966). *Models and Analogies in Science*. University of Notre Dame Press.

Hilborn, R. and M. Mangel (1997). *The Ecological Detective: Confronting Models with Data*, Volume 28. Princeton University Press.

Hitchcock, C. and E. Sober (2004). Prediction versus accommodation and the risk of overfitting. *The British Journal for the Philosophy of Science 55*(1), 1–34.

Holling, C. S. (1973). Resilience and stability of ecological systems. *Annual Review of Ecology and Systematics 4*(1), 1–23.

Hooper, D. U., E. C. Adair, B. J. Cardinale, et al. (2012). A global synthesis reveals biodiversity loss as a major driver of ecosystem change. *Nature 486*(7401), 105.

Horn, H. S. and R. H. MacArthur (1972). Competition among fugitive species in a harlequin environment. *Ecology 53*(4), 749–752.

Horwich, P. (1991). On the nature and norms of theoretical commitment. *Philosophy of Science 58*(1), 1–14.

Horwich, P. (2016). *Probability and Evidence*. Cambridge University Press.

Howard, L. O. L. O. and W. F. Fiske (1911). *The Importation into the United States of the Parasites of the Gipsy Moth and the Brown-Tail Moth: A Report of Progress, with Some Consideration of Previous and Concurrent Efforts of this Kind* no. 91. U.S. Deptartment of Agriculture, Bureau of Entomology, https://www.biodiversitylibrary.org/bibliography/65198.

Howson, C. and P. Urbach (2006). *Scientific Reasoning: The Bayesian Approach*. Open Court Publishing.

Hughes, R. I. (1997). Models and representation. *Philosophy of Science 64*, S325–S336.

Isbell, F., V. Calcagno, A. Hector, et al. (2011). High plant diversity is needed to maintain ecosystem services. *Nature 477*(7363), 199.

James, W. (1979). The Will to Believe and Other Essays in Popular Philosophy, Volume 6. Harvard University Press.

Justus, J. (2008a). Complexity, diversity, and stability. In S. Sahotra and A. Plutynski (Eds.), *A Companion to the Philosophy of Biology*, 321–350. John Wiley & Sons.

Justus, J. (2008b). Ecological and Lyapunov stability. *Philosophy of Science 75*(4), 421–436.

Justus, J. (2011). A case study in concept determination: Ecological diversity. *Handbook of the Philosophy of Ecology*, pp. 147–168. Elsevier.

Kellert, S. H. (1994). *In the Wake of Chaos: Unpredictable Order in Dynamical Systems*. University of Chicago Press.

Kendall, B. E., C. J. Briggs, W. W. Murdoch, et al. (1999). Why do populations cycle? A synthesis of statistical and mechanistic modeling approaches. *Ecology 80*(6), 1789–1805.

Kingsland, S. E. (1995). *Modeling Nature*. University of Chicago Press.

Kitcher, P. (1989). Explanatory unification and the causal structure of the world. In P. Kitcher and W. Salmon (Eds.), *Scientific Explanation*, pp. 410–505. University of Minnesota Press.

Kitcher, P. (1995). *The Advancement of Science: Science without Legend, Objectivity without Illusions*. Oxford University Press.

Kricher, J. (2009). *The Balance of Nature: Ecology's Enduring Myth*. Princeton University Press.

Kuhn, T. S. (1977). *The Essential Tension: Selected Studies in Scientific Teadition and Change*. University of Chicago Press.

Kulvicki, J. V. (2013). *Images*. Routledge.

Lande, R. (1987). Extinction thresholds in demographic models of territorial populations. *The American Naturalist 130*(4), 624–635.

Lande, R. (1988a). Demographic models of the northern spotted owl (*Strix occidentalis caurina*). *Oecologia 75*(4), 601–607.

Lande, R. (1988b). Genetics and demography in biological conservation. *Science 241*(4872), 1455–1460.

Lange, M. (2005). Ecological laws: What would they be and why would they matter? *Oikos 110*(2), 394–403.

Lasersohn, P. (1999). Pragmatic halos. *Language 75*(3), 522–551.

Laudan, L. (1986). *Science and Values: The Aims of Science and Their Role in Scientific Debate*. Univ of California Press.

Lawlor, L. (1978). A comment on randomly constructed model ecosystems. *The American Naturalist 112*(984), 445–447.

Lawton, J. H. (1999). Are there general laws in ecology? *Oikos 84*(2), 177–192.

Levi, I. (1967). *Gambling with Truth: An Essay on Induction and the Aims of Science* MIT Press.

Levins, R. (1966). The strategy of model building in population biology. *American Scientist 54*(4), 421–431.

Levins, R. (1969). Some demographic and genetic consequences of environmental heterogeneity for biological control. *American Entomologist 15*(3), 237–240.

Levins, R. and D. Culver (1971). Regional coexistence of species and competition between rare species. *Proceedings of the National Academy of Sciences 68*(6), 1246–1248.

Linquist, S., T. R. Gregory, T. A. Elliott, *et al.* (2016). Yes! There are resilient generalizations (or "laws") in ecology. *The Quarterly Review of Biology 91*(2), 119–131.

Lloyd, E. A. (1994). *The Structure and Confirmation of Evolutionary Theory*. Princeton University Press.

Loreau, M., S. Naeem, P. Inchausti, et al. (2001). Biodiversity and ecosystem functioning: Current knowledge and future challenges. *Science 294*(5543), 804–808.

MacArthur, R. H. and E. O. Wilson (1967). *The Theory of Island Biogeography*. Princeton University Press.

Maher, P. (1993). *Betting on Theories*. Cambridge University Press.

Marquard, E., A. Weigelt, V. M. Temperton, et al. (2009). Plant species richness and functional composition drive overyielding in a six-year grassland experiment. *Ecology 90*(12), 3290–3302.

Matthewson, J. (2011). Trade-offs in model-building: A more target-oriented approach. *Studies in History and Philosophy of Science Part A 42*(2), 324–333.

Matthewson, J. and M. Weisberg (2009). The structure of tradeoffs in model building. *Synthese 170*(1), 169–190.

May, R. (1973a). Stability and complexity in model ecosystems. *Monographs in Population Biology 6*, 1.

May, R., J. Lawton, and N. Stork (1995). Assessing extinction rates. In R. May and J. Lawton (Eds.), *Extinction Rates*, Chapter 1, pp. 1 – 24. Oxford University Press.

May, R. M. (1973b). Qualitative stability in model ecosystems. *Ecology 54*(3), 638–641.

May, R. M. (1974). Biological populations with nonoverlapping generations: Stable points, stable cycles, and chaos. *Science 186*(4164), 645–647.

May, R. M. (1975). Biological populations obeying difference equations: Stable points, stable cycles, and chaos. *Journal of Theoretical Biology 51*(2), 511–524.

May, R. M. (1976). Simple mathematical models with very complicated dynamics. *Nature 261*(5560), 459.

May, R. M. (1990). How many species? *Philosophical Transactions of the Royal Society of London B 330*(1257), 293–304.

May, R. M. (2002). The best possible time to be alive: The logistic map, In G. Farmelo (Ed.), *It Must Be Beautiful: Great Equations of modern Science*, pp. 28–45. Granta Books.

May, R. M. (2011). Why worry about how many species and their loss? *PLoS Biology 9*(8), e1001130.

May, R. M. and G. F. Oster (1976). Bifurcations and dynamic complexity in simple ecological models. *The American Naturalist 110*(974), 573–599.

McCann, K. (2005). Perspectives on diversity, structure, and stability. In K. Cuddington and B. Beisner (Eds.), *Ecological Paradigms Lost: Routes of Theory Change*, pp. 183–200. Elsevier Academic Press.

McMullin, E. (1985). Galilean idealization. *Studies in History and Philosophy of Science Part A 16*(3), 247–273.

Mikkelson, G. M. (1997). Methods and metaphors in community ecology: The problem of defining stability. *Perspectives on Science 5*, 481–498.

Mikkelson, G. M. (2003). Ecological kinds and ecological laws. *Philosophy of Science 70*(5), 1390–1400.

Moll, R. J., D. Steel, and R. A. Montgomery (2016). Aic and the challenge of complexity: A case study from ecology. *Studies in History and Philosophy of Science Part C: Studies in History and Philosophy of Biological and Biomedical Sciences 60*, 35–43.

Mora, C., D. P. Tittensor, S. Adl, A. G. Simpson, and B. Worm (2011). How many species are there on earth and in the ocean? *PLoS Biology 9*(8), e1001127.

Morrison, M. (2015). *Reconstructing Reality: Models, Mathematics, and Simulations*. Oxford University Press.

Myers, N. (1979). *The Sinking Ark: A New Look at the Problem of Disappearing Species*. Pergamon Press.

Nicholson, A. J. (1954). An outline of the dynamics of animal populations. *Australian Journal of Zoology 2*(1), 9–65.

Nicholson, A. J. (1957). The self-adjustment of populations to change. In *Cold Spring Harbor Symposia on Quantitative Biology*, Volume 22, pp. 153–173. Cold Spring Harbor Laboratory Press.

Nicholson, A. J. and V. A. Bailey (1935). The balance of animal populations – Part I. *Journal of Zoology 105*(3), 551–598.

Norse, E. A. (1989). *Ancient Forests of the Pacific Northwest*. Island Press.

Odenbaugh, J. (2001). Ecological stability, model building, and environmental policy: A reply to some of the pessimism. *Philosophy of Science 68*(S3), S493–S505.

Odenbaugh, J. (2003). Complex systems, trade-offs, and theoretical population biology: Richard Levin's "strategy of model building in population biology" revisited. *Philosophy of Science 70*(5), 1496–1507.

Odenbaugh, J. (2005). Idealized, inaccurate but successful: A pragmatic approach to evaluating models in theoretical ecology. *Biology and Philosophy 20*(2–3), 231–255.

Odenbaugh, J. (2006b). Message in the bottle: The constraints of experimentation on model building. *Philosophy of Science 73*(5), 720–729.

Odenbaugh, J. (2006a). The strategy of "the strategy of model building in population biology." *Biology and Philosophy 21*(5), 607–621.

Odenbaugh, J. (2006c). Struggling with the science of ecology. *Biology & Philosophy 21*(3), 395.

Odenbaugh, J. (2010). Models. In S. Sarkar and A. Plutynski (Eds.), *A Companion to the Philosophy of Biology*, pp. 506–524. John Wiley & Sons.

Odenbaugh, J. (2011a). Complex ecological systems. In *Philosophy of Complex Systems*, pp. 421–439. Elsevier.

Odenbaugh, J. (2011b). True lies: Realism, robustness, and models. *Philosophy of Science 78*(5), 1177–1188.

Odenbaugh, J. (2015). Semblance or similarity? Reflections on simulation and similarity. *Biology & Philosophy 30*(2), 277–291.

Odenbaugh, J. (2018). Models, models, models: A deflationary view. *Synthese*, 1–16.

Odenbaugh, J. and A. Alexandrova (2011). Buyer beware: Robustness analyses in economics and biology. *Biology & Philosophy 26*(5), 757–771.

Orians, G. H. (1975). Diversity, stability and maturity in natural ecosystems. In *Unifying Concepts in Ecology*, pp. 139–150. Springer.

Orzack, S. H. (2005). Discussion: What, if anything, is "the strategy of model building in population biology" A comment on Levins (1966) and Odenbaugh (2003). *Philosophy of Science 72*(3), 479–485.

Orzack, S. H. and E. Sober (1993). A critical assessment of Levins's the strategy of model building in population biology (1966). *The Quarterly Review of Biology 68*(4), 533–546.

Otto, S. P. and T. Day (2011). *A Biologist's Guide to Mathematical Modeling in Ecology and Evolution*. Princeton University Press.

Paine, R. T. (1988). Road maps of interactions or grist for theoretical development? *Ecology 69*(6), 1648–1654.

Parker, W. S. (2009). Ii?Confirmation and adequacy-for-purpose in climate modelling. In *Aristotelian Society Supplementary Volume*, Volume 83, pp. 233–249. John Wiley & Sons.

Pereira, H. M., L. Borda-de Água, and I. S. Martins (2012). Geometry and scale in species–area relationships. *Nature 482*(7386), E3.

Perini, L. (2005a). The truth in pictures. *Philosophy of Science 72*(1), 262–285.

Perini, L. (2005b). Visual representations and confirmation. *Philosophy of Science 72*(5), 913–926.

Peters, R. H. (1991). *A Critique for Ecology*. Cambridge University Press.

Pimm, S. L. (1984). The complexity and stability of ecosystems. *Nature 307*(5949), 321.

Pimm, S. L. and R. A. Askins (1995). Forest losses predict bird extinctions in eastern North America. *Proceedings of the National Academy of Sciences 92*(20), 9343–9347.

Pincock, C. (2011). *Mathematics and Scientific Representation*. Oxford University Press.

Popper, K. R. (1979). *Objective Knowledge: An Evolutionary Approach*. Oxford University Press.

Potochnik, A. (2017). *Idealization and the Aims of Science*. University of Chicago Press.

Provine, W. B. (2001). *The Origins of Theoretical Population Genetics: With a New Afterword*. University of Chicago Press.

Reid, W. V. (1992). How many species will there be. *Tropical Deforestation and Species Extinction 55*, 55–57.

Renshaw, E. (1993). *Modelling Biological Populations in Space and Time*, Volume 11. Cambridge University Press.

Ricker, W. E. (1954). Stock and recruitment. *Journal of the Fisheries Board of Canada 11*(5), 559–623.

Rockwood, L. L. (2015). *Introduction to Population Ecology*. John Wiley & Sons.

Rosenzweig, M. L. (1995). *Species Diversity in Space and Time*. Cambridge University Press.

Rosenzweig, M. L. (2003). Reconciliation ecology and the future of species diversity. *Oryx 37*(2), 194–205.

Royall, R. (1997). *Statistical Evidence: A Likelihood Paradigm*. Routledge.

Rudner, R. (1953). The scientist qua scientist makes value judgments. *Philosophy of Science 20*(1), 1–6.

Salmon, W. (1984). *Scientific Explanation and the Causal Structure of the World*. Princeton University Press.

Sarkar, S. (2005). *Biodiversity and Environmental Philosophy: An Introduction*. Cambridge University Press.

Shrader-Frechette, K. S. and E. McCoy (1993). *Method in Ecology: Strategies for Conservation*. Cambridge University Press.

Simberloff, D. (1981). The sick science of ecology: Symptoms, diagnosis, and prescription. *Eidema 1*, 49–54.

Simberloff, D. (1992). Do species–area curves predict extinction in fragmented forest? In T. C. Whitmore and J. A. Sayer (Eds.). *Deforestation and Species Extinction in Tropical Moist Forests*, pp. 75–89. Chapman & Hall.

Smart, J. J. C. (1963). *Philosophy and Scientific Realism*. Routledge.

Smith, F. E. (1961). Density dependence in the Australian thrips. *Ecology 42*(2), 403–407.

Smith, H. S. (1935). Biotic factors in the determination of population densities. *Journal of Economic Entomology 28*(6), 873–898

Smith, P. (1998). *Explaining Chaos*. Cambridge University Press.

Sober, E. (1985). Constructive empiricism and the problem of aboutness. *The British Journal for the Philosophy of Science 36*(1), 11–18.

Sober, E. (1997). Two outbreaks of lawlessness in recent philosophy of biology. *Philosophy of Science 64*, S458–S467.

Sober, E. (2002). Instrumentalism, parsimony, and the akaike framework. *Philosophy of Science 69*(S3), S112–S123.

Sorensen, R. (2012). Veridical idealizations. *Thought Experiments in Philosophy, Science, and the Arts 11*, 30.

Steel, D. (2010). Epistemic values and the argument from inductive risk. *Philosophy of Science 77*(1), 14–34.

Steel, D. (2013). Acceptance, values, and inductive risk. *Philosophy of Science 80*(5), 818–828.

Sterelny, K. (1990). *The Representational Theory of Mind: An Introduction*. Basil Blackwell.

Stork, N. E. (1993). How many species are there? *Biodiversity and Conservation 2*, 215–232.

Stork, N. E. (2010). Re-assessing current extinction rates. *Biodiversity and Conservation 19*(2), 357–371.

Strevens, M. (2003). The role of the priority rule in science. *The Journal of Philosophy 100*(2), 55–79.

Strevens, M. (2008). *Depth: An Account of Scientific Explanation*. Harvard University Press.

Strevens, M. (2017). The structure of asymptotic idealization. *Synthese*, 1–19.

Suárez, M. (2010). Scientific representation. *Philosophy Compass 5*(1), 91–101.

Suárez, M. (2015). Deflationary representation, inference, and practice. *Studies in History and Philosophy of Science Part A 49*, 36–47.

Sugihara, G., R. May, H. Ye, et al. (2012). Detecting causality in complex ecosystems. *Science*, 1227079.

Suppe, F. (1989). *The Semantic Conception of Theories and Scientific Realism*. University of Illinois Press.

Suppes, P. (1957). *Introduction to Logic*. Van Nostran Reinhold Company.

Suppes, P. (1966). Models of data. In *Studies in Logic and the Foundations of Mathematics*, Volume 44, pp. 252–261. Elsevier.

Taper, M. L. and S. R. Lele (2010). *The Nature of Scientific Evidence: Statistical, Philosophical, and Empirical Considerations*. University of Chicago Press.

Teller, P. (2012). Modeling, truth, and philosophy. *Metaphilosophy 43*(3), 257–274.

Thomas, C. D. and M. Williamson (2012). Extinction and climate change. *Nature 482*(7386), E4.

Thompson, P. (1989). *The Structure of Biological Theories*. State University of New York Press.

Tilman, D. and J. A. Downing (1994). Biodiversity and stability in grasslands. *Nature 367*(6461), 363.

Tilman, D., R. M. May, C. L. Lehman, and M. A. Nowak (1994). Habitat destruction and the extinction debt. *Nature 371*(6492), 65.

Triantis, K. A., F. Guilhaumon, and R. J. Whittaker (2012). The island species–area relationship: Biology and statistics. *Journal of Biogeography 39*(2), 215–231.

Turchin, P. (2001). Does population ecology have general laws? *Oikos 94*(1), 17–26.

van Fraassen, B. C. (1980). *The Scientific Image*. Oxford University Press.

van Fraassen, B. C. (2010). *Scientific Representation: Paradoxes of Perspective*. Oxford University Press.

Weber, M. (1999). The aim and structure of ecological theory. *Philosophy of Science 66*(1), 71–93.

Weisberg, M. (2006). Robustness analysis. *Philosophy of Science 73*(5), 730–742.

Weisberg, M. (2012). *Simulation and Similarity: Using Models to Understand the World*. Oxford University Press.

Weisberg, M. and K. Reisman (2008). The robust volterra principle. *Philosophy of Science 75*(1), 106–131.

Wilson, E. O. (1988). The current state of biological diversity. In E. O. Wilson and F. M. Peter (Eds.), *Biodversity*, pp. 3–20. National Academy Press.

Wilson, E. O. (1992). *The Diversity of Life*. W. W. Norton & Company.

Wimsatt, W. C. (2007). *Re-engineering Philosophy for Limited Beings: Piecewise Approximations to Reality*. Harvard University Press.

Wojdak, J. M. and G. G. Mittelbach (2007). Consequences of niche overlap for ecosystem functioning: An experimental test with pond grazers. *Ecology 88*(8), 2072–2083.

Woodward, J. (2003). *Making Things Happen: A Theory of Causal Explanation*. Oxford University Press.

Yaffee, S. L. (1994). *The Wisdom of the Spotted Owl: Policy Lessons for a New Century*. Island Press.

Ye, H., R. J. Beamish, S. M. Glaser, et al. (2015). Equation-free mechanistic ecosystem forecasting using empirical dynamic modeling. *Proceedings of the National Academy of Sciences 112*(13), E1569–E1576.

Yodzis, P. (1981). The stability of real ecosystems. *Nature 289*(5799), 674.

Acknowledgments

Originally, I wanted to write my dissertation on the metaphysics of species. It seemed obvious. My graduate school advisor Marc Ereshefsky is an expert on the topic. Still, he persuaded me to think more adventurously; after all, what else could be written on that topic? Little did we know. He very insightfully asked me what else was of interest to me? I had been reading Sharon Kingsland's (1995) history of population ecology and William Provine's (2001) history of population genetics. I mentioned I was really perplexed about the rampant role of idealizations in biological models, but especially in ecology. What place could models known to be false have in biology? Additionally, very few philosophers had mined ecology philosophically. We did an intensive directed study on models to see whether it would fly. With his encouragement, I buried myself in the philosophical literature while taking courses in population, community ecology, and theoretical ecology. Thereafter, I had a dissertation topic, and, ultimately, a dissertation. Without Marc, you wouldn't be reading this.

Once my dissertation was being written, I needed to find a committee. I called a renowned population ecologist Ed McCauley at the University of Calgary. I told him of my project, and his one question to me was: "You know theoretical ecology has *lot* of math?" Gulp. I said a hesitant "yes," and he agreed. He allowed me to take his courses and hangout in his lab provided I was willing to do the work. I was, and so I did. The next step was finding a philosopher of biology for whom model building was their forte, and that was Bill Wimsatt. Bill was not only my external committee member, but he also supported me and my work after graduate school. I am very grateful to Bill for his generosity. In many respects, I have been a worker in his research program.

Over the years, I have accumulated many debts. My thanks go to those who have helped me and my work, which includes André Ariew, William Bausman, Robert Brandon, Greg Cooper, Rebecca Copenhaver, Elizabeth Crone, Adrian Currie, Kevin deLaplante, Michael Dietrich, Steve Downes, Chris Eliot, Alkistis Elliott-Graves, Marc Ereshefsky, Nicholas Gotelli, Paul Griffiths, Jim Griesemer, Cliff Hooker, James Justus, Richard Levins, Stefan Linquist, Lisa Lloyd, John Matthewson, Greg Mikkelson, Elaine Landry, Ed McCauley, Roberta Millstein, Bill Nelson, Steve Orzack, Anya Plutynski, Grant Ramsey, Collin Rice, Alex Rosenberg, Bill Rottschaefer, Alex Rueger, Michael Ruse, Sahotra Sarkar, Elliott Sober, Alex von Stein, Kim Sterelny, Peter Taylor, Dennis Walsh, Ken Waters, Michael Weisberg, Bill Wimsatt, and Jim Woodward.

I thank my family for being so supportive over these many years. I especially thank Stacy who loves science at least as much as I do. She is selflessly supportive of me and the work I love. And, finally, I thank my son Everett who was born as the book was being finished. I hope he loves science too, but it is okay if he doesn't.

Philosophy of Biology

Grant Ramsey

KU Leuven

Grant Ramsey is a BOFZAP Research Professor at the Institute of Philosophy, KU Leuven, Belgium. His work centers on philosophical problems at the foundation of evolutionary biology. He has been awarded the Popper Prize twice for his work in this area. He also publishes in the philosophy of animal behavior, human nature, and the moral emotions. He runs the Ramsey Lab (theramseylab.org), a highly collaborative research group focused on issues in the philosophy of the life sciences.

Michael Ruse

Florida State University

Michael Ruse is the Lucyle T. Werkmeister Professor of Philosophy and the Director of the Program in the History and Philosophy of Science at Florida State University. He is Professor Emeritus at the University of Guelph, in Ontario, Canada. He is a former Guggenheim fellow and Gifford lecturer. He is the author or editor of over sixty books, most recently *Darwinism as Religion: What Literature Tells Us about Evolution*; *On Purpose*; *The Problem of War: Darwinism, Christianity, and their Battle to Understand Human Conflict*; and *A Meaning to Life*.

About the Series

This Cambridge Elements series provides concise and structured introductions to all of the central topics in the philosophy of biology. Contributors to the series are cutting-edge researchers who offer balanced, comprehensive coverage of multiple perspectives, while also developing new ideas and arguments from a unique viewpoint.

Cambridge Elements ☰

Philosophy of Biology

Printed in the United States
By Bookmasters